A Sioux Chronicle

The Civilization of the American Indian Series

GEORGE E. HYDE

A Sioux Chronicle

Norman

UNIVERSITY OF OKLAHOMA PRESS

By George E. Hyde

Red Cloud's Folk: A History of the Oglala Sioux Indians (Norman, 1937; revised edition, 1957)
The Pawnee Indians (Denver, 1951; Norman, 1974)
Rangers & Regulars (Columbus, 1953)
A Sioux Chronicle (Norman, 1956)
Indians of the High Plains: From the Prehistoric Period to the Coming of Europeans (Norman, 1959)
Spotted Tail's Folk: A History of the Brulé Sioux (Norman, 1961)
Indians of the Woodlands: From Prehistoric Times to 1725 (Norman, 1962)
Life of George Bent: Written from His Letters (Norman, 1968)

Library of Congress Catalog Card Number: 56–11233

A Sioux Chronicle is volume 45 in *The Civilization of American Indian Series.*

TO MY SISTER, MABEL L. REED,

I DEDICATE THIS

SECOND BOOK OF SIOUX HISTORY

Preface

THE PRESENT VOLUME is a continuation of the history of the Sioux as it was set down in *Red Cloud's Folk*, an earlier book which dealt with the history of this tribe up to the end of the Sioux war of 1876–77. In *Red Cloud's Folk* the Sioux were depicted as a free and roving race of hunters and warriors, living in moving camps and gaining a livelihood by hunting buffalo and plundering neighbor tribes. In the present book the Sioux are shown as a captive people confined to a reservation in Dakota, their means of making a living in their old way destroyed by the whites and their tribe being fed, clothed, cared for and controlled by agents of the United States government. Unconquered and still defiant, the Sioux bitterly resented the government's policy of confining them to a restricted area and of forcing them to comply with certain regulations which meant that they must alter their old customs and rapidly adjust themselves to new conditions; and from 1878 to 1890 a struggle was in progress on the great Sioux reservation, with the Indians striving to cling to the free ways of their ancestors while the government officials were working systematically to break the tribe to pieces, to destroy the chiefs, and to compel the Sioux to accept white leadership.

It has been said that Sioux history ended with the surrender of Crazy Horse and the flight of Sitting Bull and his die-hards to Canada in 1877, and that might be true if history was nothing more than a record of wars and battles. Such a view, however, would ignore the fact that it was after the end of the war of 1876–77 that the Sioux, forced to concentrate on their great reservation in Dakota, were for the first time really united and that during the years from 1878 to 1890 they had to fight a peaceful war against the whole power of the United States government, which was striving to dominate them. The government officials, and the leaders of the Christian and benevolent associations that were meddling in Indian affairs, termed this policy "Progress in Civilization"; but it was in fact a forcing process concocted by social theorists who were in such haste to make the Sioux over into imitations of white people that they overlooked the fact that an Indian tribe might be ruled rather than destroyed and that if it were really necessary to eradicate ancient Indian customs, it might be better to do this gradually rather than by the employment of sudden force. Young Theodore Roosevelt spoke his mind about this in 1892, when he told a meeting of self-termed Indian Friends that if they expected to cause these very backward tribes to skip suddenly over six thousand years of life and experience and come suddenly up level with the white race, they were going to be sadly disappointed.

The writing of Sioux history seems always to end with 1877, though some accounts of the ghost dance troubles of 1890 have been published. In the present work the history of events from 1878 to 1891 is set down in detail. Here are unknown chapters in the lives of Spotted Tail, Red Cloud, and the other chiefs. Here we behold Spotted Tail blandly pushing his tame government agent aside and running his tribal affairs in his own manner; Red Cloud engaged in a furious five years' war with an agent who was far from tame; great government farming programs developed for Indians who had no intention of farming; and land councils supposedly too dull to be put into a book but dramatized by mobs of enraged Sioux flourishing Winchester rifles and by the trickery and bribery of land commissions and their agents. There is no war from 1877 on; but nearly every year there is a crisis that seems to threaten open conflict, and in 1890 war does come —a rather tame war, but serious enough to bring practically the whole United States Army into the Sioux land.

The author makes no apology for criticizing the Indian policies of the period 1877–90 or for speaking plainly concerning the men who evolved these policies and attempted to carry them out. In the main, these policies were invented by visionaries who knew practically nothing about wild Indians but who regarded their own good intentions as a sufficient justification for meddling in other people's lives. These visionaries had complete confidence in their ability to push the Sioux suddenly onward and upward from barbarism into Christianity and self-support from agriculture in a few years' time. At first they pledged themselves to have all the Sioux at work farming in four to five years (by 1872), but by 1877, with the Indians still stubbornly resisting all efforts to compel them to work, the crusaders ceased to name a date for the accomplishment of their plans. They gave a pledge, however, to Congress and the country that they would succeed, perhaps in ten years or a little longer. Urged on by these crusaders, the government spent untold millions of dollars on the Sioux farming experiment, which failed, was revived, and failed again. It was not only Sioux apathy but Dakota drought that caused these failures. Even the Sioux who could be induced to try farming were completely discouraged when drought came down and destroyed their little crops.

In 1870 the Sioux were still in the hunter stage. Primitive people naturally drift from hunting to a pastoral life, and then by slow evolution take to tilling the soil. While the policy makers were intent on attempting to force the Sioux to farm, the Indians were slowly drifting into the business of cattle growing. By 1914 they had large herds and were selling beef cattle in considerable numbers. Their cattle dealing was controlled by government agents. A Sioux had to obtain the permission of the agent before he could sell his beeves; the agent dictated the terms of the sale and impounded the money the Indian received. This system was necessary, as most of the Sioux fullbloods lacked the shrewdness of white men. They were like children, ready to sell their cattle at any price when a whim to buy something took them, and spending their money for useless articles that took their fancy.

Toward the end of World War I the price of beef went very high, and certain white worthies began to cast covetous eyes on the Sioux herds. By some method that has never been clearly explained, the government rules were set aside and the Indians were freed from restrictions. They at once set about selling off their cattle. For a few years they lived lavishly, spending on every whim. Salesmen tempted them, and the reservation was soon invaded by fleets of new and shiny Fords. There were no real roads; no Sioux took the trouble to build a shelter for his car, and he knew nothing about machines. Presently the reservation was strewn with derelict cars. The Indians sold them as scrap. They had now sold nearly all their cattle and many families were facing starvation.

Fifty years had now passed since the Sioux had left their hunting camps in the buffalo plains to come to the reservation. They had not done badly in their slow advance toward civilization. Hampered in every way by foolish government policies, they had advanced in little more than one generation from the condition of fierce warriors to that of peaceful cattle herders. If they had lost their herds, that was the fault of government agents who went to sleep or turned their backs while white men talked the Sioux into selling. It was also the fault of the officials that the Indians, having been induced to sell their cattle, leased their lands to white men and sat down to live on their rents. A picture of conditions on the Sioux reservation at this time is enlightening. At Red Cloud's reservation—Pine Ridge—in 1920 there were 7,267 Indians who owned 2,525,378 acres of land, nearly all of which was leased for a pittance to white cattlemen. The Indians had only 2,500 cattle left. There was almost no farming by Sioux fullbloods. Education, which was started in 1879 and was to work marvels for the Sioux, was in a bad way and was accomplishing very little. After nearly fifty years of schooling most Sioux fullbloods had difficulty in understanding and speaking English. When the cattle were sold off, free rations had to be resumed to prevent suffering. Half the Pine Ridge Indians were put on rations; but by 1920 a government effort to cut down had reduced the ration roll at this reservation to 700. Eighty-five per cent of the Pine Ridge Indian families lived in old one-room houses, largely ancient and dilapidated log cabins. Most of the homes did not even have wooden floors. There were 223 graduates of the big off-the-reservation Indian training schools at Pine Ridge. These boys had been given from four to six years' training at carpentry and other trades; yet none of them seemed to be competent to do the simple work of fixing a leak in a roof. Obviously there was something very much wrong with the costly system of taking Indian boys away to distant schools and teaching them trades. They did not learn the trades they were taught. Here again the flaw in the government plan was probably the same as in all its plans—forcing. It attempted to force the Sioux to farm, to learn English, to learn trades. In every instance it failed.

Going their own way and at their own pace seemed to be the one effective manner in which the Sioux could accomplish anything. Early in the 1880's the Pine Ridge Indians had hit on a method for making a living that suited them well. Large numbers of these Indians went off with Wild West shows and circuses. They traveled all over the country, mingling with the whites and learning their ways. They went to Europe and toured England, France, Spain, Italy, and Germany with Wild West shows. Their work was to pose as warriors, and it suited them to keep up the ways of warriors and the old pagan Sioux practices. Christianity and education meant nothing to them; and coming hame to Pine Ridge, they leavened the lump of the stay-at-home Sioux, teaching them to look down on the tame tasks of farming and the childish teachings of the little Indian day schools. These much traveled Sioux had learned English effectively. It was not at all the orthodox English the schools were striving to drum into the Sioux children's heads; it was a wild growth, cowboy English, but very effective.

The conditions at all the Sioux reservations in Dakota in 1920 were about the same as at Pine Ridge. At Rosebud, east of Pine Ridge, the Spotted Tail tribe of Brûlés numbered 5,466 and had 1,867,706 acres of land, nearly all leased to white cattlemen and farmers. The Sioux here, as at Pine Ridge, had sold off their cattle and were living on lease money in the main. They had gone on free rations after selling their cattle; but by 1920 most of them were off the ration roll again. Housing was the same as at Pine Ridge—nearly all dwellings were one room; many, old log cabins. Of 1,700 Indian homes, 476 had wooden floors. Indian farming was practically nonexistent and education was in a sad way. Regarding schools as a form of government folly, the Sioux of Rosebud were very angry because the government was dipping into their tribal funds and giving Sioux money to the Roman Catholic boarding school on their reservation. The Sioux attitude was that if the whites wished to waste money on schools, they should waste their own money.

This was the condition among the Sioux in 1920. There was almost no farming; the cattle had been sold off. The Indians were living on a little money obtained by leasing their lands to white men and on government aid. The Sioux had a terribly low living standard, so low that it shocked any intelligent person who looked into matters on the reservation. About 1925 the officials were shamed into doing something about this; but they had nothing new in mind, and so they resurrected the original idea of the crusaders of 1868 and decided to make the Sioux suddenly self-supporting through farming. They planned this with the Sioux clearly as averse to farming as ever and with a cycle of Dakota drought due shortly. With Congress unwilling to vote large sums for this fresh experiment, the officials dipped into the Sioux tribal funds for the necessary means. Officials made a blueprint of a "five-year plan," employing experts from agricultural colleges to draw up plans and to supervise the work. Batteries of expensive farm machinery were bought; white "boss farmers" were employed to drive the unwilling Sioux to labor; and more acres were plowed and planted than ever before. Bright reports of rapid progress were issued. Then came silence. Then a report that the Sioux were eating ponies.

No one who knew these Indians needed further information. Pony eating meant the Sioux were starving. Gradually the truth seeped out of the reservation. Drought had taken all the carefully planted crops. The government funds and the Sioux money were gone, and only the Dakota drought remained.

Then the great depression struck and the New Deal took over. Franklin D. Roosevelt's bright young men were loaded with optimism, but after a survey of drought conditions on the Sioux reservation even they could not feel justified in spending more public money on farming. They fed and cared for the Sioux and put them to work on projects, most of which are now gone and forgotten. The Sioux, however, lived well under the New Deal and had a lot of fun out of such schemes as the Pine Ridge Buffalo Pasture and the Turkey Ranch. Then Pearl Harbor put a sudden stop to all this. Congress took the Indian funds for war uses, and the Sioux were presently told to go away somewhere and try to find work for themselves. Many of the young men were in the armed services; large numbers of Sioux left the reservation, taking their families, and found work on farms or in towns. Thus through the accident of war the Sioux were for the first time freed from the blighting controls of life on a reservation and left to shift for themselves, and on the whole they did well.

Today there are little colonies of Sioux in many towns: in Rapid City, Sioux Falls, Sioux City, Omaha, and Denver. These Indians are doing as well as anyone has a right to expect. They are learning to stand on their feet; but they feel safer by knowing that the reservation is still there, and that they can return to it as to a safe refuge in time of need. That is where the old people, the infirm, and the stay-at-homes are, and that is the home to which the Sioux wish to return from time to time. The planners who are talking of abolishing reservations might take these facts into consideration. The Sioux have paid for these reservations literally with blood, sweat, and tears, and they should be permitted to keep these last tribal homes as refuges for their people in time of stress. The reservations as places where the tribe can be made self-supporting are dead as Pharaoh; but the Indians love these lands and should be permitted to retain them.

In the preparation of this volume the usual official sources have been used and contemporary newspapers consulted, but much of the material was obtained from Indians and whites in Dakota, some of whom were witnesses of the events recorded. Over a period of years much aid was given by the late Philip F. Wells of Pine Ridge and the late C. I. Leedy of Rapid City. Tom Wells, now dead, and his wife, Flora G. Wells, were always friendly and helpful, as were also the late John Colhoff of Pine Ridge and his close friend Joseph Eagle Hawk. From Rosebud, material was obtained from such men as David A. Whipple, an employee at

Spotted Tail Agency from 1871 on, and the old Sioux men, Knife Scabbard, Sore Eyes, David Murray, and others. Will G. Robinson, the head of the South Dakota Historical Society, has kindly supplied much original material and some old photographs. More recently, much interesting material from old Dakota newspapers was supplied by Dean Herbert S. Schell of the Graduate School, South Dakota University, and by Dr. Everett W. Sterling of the College of Arts and Sciences of the same institution. Harry Anderson, a young graduate of this university, has given invaluable aid in the form of notes from old newspapers and material taken from the National Archives in Washington. In closing this preface, the author wishes to express again his thanks to all friends who have assisted him in this work.

Contents

Illustrations

Maps

A Sioux Chronicle

I

The Great Reservation

AT THE close of the Indian war of 1876–77 the Western or Teton Sioux had been deprived of the Black Hills and of the vast hunting grounds in the Powder River, Tongue River, and Bighorn country; but they still held by treaty right a great reservation in Dakota, extending from the Missouri River west to the 103rd meridian, just east of the Black Hills, and from the Nebraska line north to the Cannonball River, a little south of the town of Bismarck in North Dakota. The reservation included some 35,000 square miles, and the Teton Sioux, estimated to number about 20,000, held the right of occupation of these lands in perpetuity, with all white settlers barred from the reservation. But, where Indians were concerned, treaty pledges were not very sure titles, and within a few months after the end of the Sioux war some members of Congress and the officials at the Indian Office were vigorously pressing a plan for the removal of all the Western Sioux to Indian Territory and the opening of the great Sioux reservation to settlement.

The Sioux had few friends among the whites in 1877. The public was still shocked over what was termed the Custer massacre; General Phil Sheridan, commanding the troops in the Sioux country, was storming at the Indians and making violent threats; and Congress was in an unsympathetic mood, seemingly only interested in making the Sioux pay dearly for daring to fight back when the army was sent to attack them in 1876. The Eastern groups of self-styled Indian Friends (assorted idealists and humanitarians) were so overawed by public anger at the Sioux that they hardly dared to speak out, but—unable to keep from meddling in some manner—many of their leaders were backing the wildcat scheme for removing all the Sioux to Indian Territory, quite unaware that in that plan lay the seeds of the greatest Indian war the country had ever experienced. The Sioux were not a meek people, and they would not have sat in their camps in the hot country of Indian Territory and quietly watched their children, women and old people die of fever. They would have set the prairie on fire, from Texas to the Canadian border.

Congress had voted for a middle course. The Sioux were to be concentrated at agencies along the west bank of the Missouri River, where they could be easily supplied with rations and other needs and where the troops could control them. Most of the Teton Sioux were already at agencies on the Missouri, but Red Cloud's Oglalas and Spotted Tail's Brûlés were at two agencies on the head of White River near the extreme southwest corner of the Sioux reservation. Indeed, both agencies were a few miles outside the reservation border, in the state of Nebraska. Assuming that these two Sioux groups would move promptly to the Missouri when asked to do so, the Indian Office had taken over the old Ponca Indian Agency, on the west bank of the Missouri in northern Nebraska, to be the new Spotted Tail Agency, and had built a new Red Cloud Agency on the west bank of the river at the mouth of Medicine Creek, not far south of the present town of Pierre, South Dakota.[1] At these new agencies all rations, clothing, and other supplies for the Sioux were being unloaded from steamboats, but at the old agencies far to the west the Red Cloud and Spotted Tail Sioux were sitting sullenly in their camps, refusing to move. Even though told that winter was coming on, that no rations or clothing were being delivered at their old agencies, and that they would soon be starving and freezing, they sat in their camps and would not budge. Told what a wonderful land the district along

[1] The little and friendly Ponca tribe was dragooned into giving up their reservation on the Missouri, north of the Niobrara River. The Poncas were sent to Indian Territory, where many of them perished of fever and hunger. The Dakota politicians and businessmen were active in this scheme to remove Red Cloud and Spotted Tail to the Missouri, for the coming of these masses of Sioux meant great profits to Yankton and other towns along the Missouri. At first it was proposed to settle Red Cloud's folk at the old Ponca reservation, but later these lands were given to Spotted Tail's Brûlés. The contract for building the new Red Cloud Agency was let as early as August, 1877, long before the Sioux had agreed to move to the Missouri.

the Missouri was, they jeered. That river land was low country, a sickly place, where their children and old people had died at the old Whetstone Agency in 1868–70. They were high country Sioux, and nothing would induce them to go and live near the detested Missouri. Besides, they did not trust the government officials. There was some trick in this, and—having enticed them to the Missouri—the whites would try to lead them into some trap.

In this moment of crisis the surrendered Oglala war chief, Crazy Horse, did not help his people particularly by forming a plan to break his pledge to General George Crook that he would remain quietly at Red Cloud Agency. Crazy Horse was a fighting Indian who felt uneasy and unsafe among the whites on the reservation, and he was planning to break away with his camp of followers and return to the Bighorn country. The fact that his band had been partly disarmed and dismounted did not deter him, but when the moment for action came, nearly all of his own followers deserted him in fear that he would lead them to their deaths and they fled to the shelter of the friendly Sioux camps at the two agencies. Mortally wounded while resisting arrest, Crazy Horse died at Fort Robinson near Red Cloud Agency, leaving behind him a legacy of hate and suspicion that for the moment bade fair to cause a resumption of hostilities between the Sioux and the troops at Fort Robinson. But the friendly agency chiefs kept a tight control over the warriors who sought to fight and, with quiet restored, the Sioux began to realize that there were more important matters for them to deal with than the seeking of vengeance for the death of Crazy Horse. The government meant just what it said: the Sioux must move to the Missouri River or face starvation the coming winter.

Still the stubborn and suspicious Indians would not give in. Red Cloud, Spotted Tail, and the other chiefs and leading men were now invited to Washington, to confer with the new President, Rutherford B. Hayes, and his secretary of the interior. The chiefs found the new Great Father a sympathetic and friendly listener. He promised them that if they would persuade their people to move to the Missouri, where rations and clothing for the winter were piled up waiting to be issued, in the following spring he would send a commission to the Sioux reservation and Red Cloud and Spotted Tail could then select any points inside the reservation that suited their people, and new agencies would be built at the points selected. The chiefs returned from Washington with printed copies of the council with the officials. In the copies they had carefully marked with colored pencils the promise the President had made. The returning chiefs told the Sioux, "One winter near the Missouri, then back to our own high country, to live permanently at places of our own selection." The Sioux held councils and decided to move. Red Cloud gathered all his camps at the old agency near Fort Robinson; Spotted Tail, on Beaver Creek to the southeast of Red Cloud, got his Brûlés ready; and the Sioux set out on the great march. Red Cloud's folk moved along the land south of White River, Spotted Tail's column headed eastward along a trail to the south of Red Cloud's. Each great moving column was accompanied by a long train of wagons, hauling rations and supplies; a herd of beef cattle went with each column; and small detachments of cavalry accompanied the march. According to Agent James Irwin, who was with Red Cloud, the Crazy Horse camp of surrendered hostiles broke away from Spotted Tail's column at a point near the head of Wounded Knee Creek and fled northward, bearing the body of their dead chief with them. They came down on Red Cloud's people, creating great excitement; but Agent Irwin avoided serious trouble by giving the runaways a quantity of rations and other supplies, and they went on their way northwest, heading for Powder River and for Sitting Bull's hostile camp in Canada.[2]

[2] Joseph Eagle Hawk, who was a small boy in the Crazy Horse camp in 1877, denies the report of Agent Irwin. He states that the Crazy Horse people were with Red Cloud during the whole march, and that it was at a camp near the mouth of South Fork of White River in midwinter, 1877–78, that the decision to go north was made. Led by Big Road, the camp slipped away, and in the end they reached Sitting Bull's camp in Canada. I still think that Irwin's report was correct. Eagle Hawk was probably with a small group of the Crazy Horse camp that had joined Red Cloud; but most of the group were with Spotted Tail. Lieutenant Jesse M. Lee was acting as agent for Spotted Tail during the march to the Missouri. When he left the Spotted Tail group in July, 1878, he told the reporter for a Yankton newspaper that when the move to the Missouri was planned only part of the Spotted Tail group approved; part refused to go to the Missouri and "deserted" and "have not since returned." This seems to refer to the break away of the Crazy Horse camp. See Lee interview, reprinted in

The Sioux of the Oglala and Brûlé groups hated the Missouri River country so much that even at this last moment they could not be induced to go all the way to their new agencies on the river. Red Cloud formed his winter camps near the forks of White River, seventy-five miles west of his agency; Spotted Tail chose a winter camp on the South Fork of White River, at or close to the point where his people later settled in the Rosebud Creek district. All that winter rations and supplies had to be carried from the agencies on the Missouri to these distant Indian camps. However, the Sioux did most of the work, packing rations and other needed supplies to their camps on ponies. Thus the new Red Cloud Agency on the Missouri, built at great expense, was a total loss. Red Cloud's folk would not camp within seventy-five miles of it. Spotted Tail was living at a point eighty miles from his agency—the old Ponca Agency. But there was Chief Swift Bear and his Corn-Loafer band, who were rated by the officials as "progressives." These were the people of the old Corn band of the Brûlés, who had planted corn on White River in the old days but had been forced by the other Brûlés to give up that (in Sioux eyes) disgraceful practice, and the Laramie Loafer band, who had lived at the military post for so long that they had developed a desire to imitate the white people. When the Brûlés were given the old Ponca Agency on the Missouri, these Corn and Loafer people actually went to the agency and settled there; and, to the rage of the proud Brûlés, these renegades under Swift Bear were now planning to start farming near Ponca Agency. The wild fellows in Spotted Tail's camp honestly believed that it was a betrayal of the tribe's honor for any Brûlé to attempt to make a living like a white man by scratching the ground—as they contemptuously termed the operation of farming. The Sioux were hunters and warriors, and they must not demean themselves by working with their hands. As to these renegades at Ponca Agency—Swift Bear and his Corn Keepers, the Loafers, Chief Milk's little camp, and those half

Niobrara, Neb., *Pioneer,* July 19, 1878. More of the Crazy Horse hostiles were reported to have run away in July, 1878, and another group, led by Short Bull who was a ghost dance leader in 1890, ran away after the Sun Dance at Rosebud in June, 1879, but they were captured on the Yellowstone.

bloods and white squawmen who had married Brûlé women—if they were only nearer to Spotted Tail's camps the tribal soldiers could whip them soundly and cure them of this disgraceful wish to work like white men. But they were too far away for the soldier-police to deal with them, and the angry warriors in Spotted Tail's camps could only fume and make threats. Old Little Thunder, formerly head chief of the Brûlés, was there at Ponca Agency, actually living with the Corn-Loafer people.[3]

Red Cloud and Spotted Tail had a hard time in the winter of 1877–78. Their people could not understand this move toward the Missouri. Why leave their own high country and move toward the sickly low lands near the river, only to move back to their own country in the spring? They could not comprehend that it cost the government over one dollar per hundred pounds per hundred miles to haul rations and other supplies to the old Red Cloud and Spotted Tail agencies in northwestern Nebraska and that Congress wished to save this expense by placing the new agencies on or near the Missouri, where they could be cheaply supplied by water transportation up the river. They felt no gratitude toward President Hayes and his secretary of the interior, who had promised to sacrifice this plan for saving public money if the Sioux would go to the Missouri for one winter, where their rations and supplies were piled up, waiting for them. The common Indians thought the Great Father should set aside the will of Congress, load the rations, clothing, and other supplies on wagons at the Missouri River agencies, and have them hauled nearly two hundred miles westward to the old agencies in Nebraska. The fact that such an operation would cost around five dollars per hundred pounds was something that did not interest the Sioux.

They had said that the country near the Missouri was a sickly district, and indeed the camps that winter were filled with sick people. Consumption cases were very bad; the adults contracted

[3] The Sioux today have forgotten Little Thunder. I have two statements concerning his last days. He is said to have died at Ponca about 1879, and Swift Bear who died years later was buried near Little Thunder, the graves being close to the site of Swift Bear's log cabin on Hay Creek. Other Sioux state that Little Thunder died in Chief Milk's camp, south of the later town of Herrick, and that he was buried there.

pneumonia; many children were down with diphtheria; and there was a shocking number of deaths. All this was the fault of the white officials who had insisted on forcing the Sioux to leave their comfortable old agencies.

If the Sioux had known what was being said in Washington, they would have been much more upset than they were. Congress was still angry with these Indians, who were regarded as responsible for the costly Sioux war of 1876–77, and the promise that President Hayes had made to the chiefs, that in the spring they could choose any location inside the Sioux reservation as the site of their new agencies, was regarded by many in Washington as a flouting of the will of Congress. That august body was half minded to set aside the President's promise and order the Sioux to be held permanently at agencies on the Missouri. The talk of removing the whole Sioux nation to Indian Territory was being kept up. No one who has written about this tribe has ever realized how near the Sioux came in 1877–78 to being uprooted and sent to Indian Territory. Even if the tribe escaped from such a disastrous decision, there was another proposal before Congress that might have affected the Sioux almost as fatally as removal to the south. That was the plan to put the tribe permanently under military control. General Sheridan was still furious with the Sioux, and he and other army officers spoke bluntly for military control. Many members of Congress backed up this proposal. Their theory was that the Indians were incurable savages who could not be taught to support themselves by civilized methods, such as farming, and that the only sensible method for dealing with people like the Sioux was to confine them closely on reservations under strict army control and keep them there until their numbers dwindled away and they became extinct. On General Sheridan's demand, the army had been given control of the Sioux agencies in 1876 as a war measure, and now the advocates of military control were demanding that the army be left in permanent charge of the Indians.

When the Sioux chiefs went to Washington, in the autumn of 1877, to confer with President Hayes, they had been considerably amused to find among the big chiefs with the Great Father a curious hairy-faced man with great owl eyes peering out through big spectacles. They called him Owl and made jokes about him. He was Carl Schurz, the new secretary of the interior, a life-long liberal and a veteran crusader for the rights of oppressed minority groups. Now, in the winter of 1877–78, while the Sioux sat in their camps on White River, complaining angrily of the way they were being treated, this queer-looking German did battle for them in Washington and saved them. Schurz fought valiantly. As a liberal he was horrified at the cold-blooded dictum of the advocates of army control, that the Indian was an incurable savage and that the only thing that could be done was to put him under army guard and wait for him and his people to die. Schurz knew almost nothing about Indians; but he had the leaders of the self-styled Indian Friends groups to instruct and advise him. In the end he carried the day by his vehement pleading for the Indians in the name of humanity and by making a personal pledge that if the Sioux were turned back to the control of the Interior Department, he would see to it that they made satisfactory progress toward self-support by farming within the next few years.

Carl Schurz unwittingly did the Sioux further yeoman service by removing "for cause" the commissioner of Indian affairs, John Q. Smith. Smith was a leading spirit in the great plan for sending the Sioux to Indian Territory, and his removal from office was the death blow to that scheme. Schurz chose as Smith's successor Ezra A. Hayt of New York state, a worthy humanitarian who had long occupied himself by dabbling in Indian welfare projects. Hayt, imagining himself to be now in complete control of Indian policy making, was happy, but only briefly. Secretary Schurz—a born crusader and overflowing with energy—had a new broom of his own, and could not imagine a better place to use it than in Hayt's domain. Almost at once he uncovered, as he thought, an "Indian Ring" in the sacred precincts of the Indian Office. Sweeping vigorously, Schurz raised such a terrible dust that poor Hayt found his own office unbearable and gladly accepted the suggestion that he should go to Dakota with the new Sioux commission, leaving Secretary Schurz to deal with accumulated dust at the Indian Office.

The Crook Treaty Council held on May 4, 1889, at the Rosebud Reservation. Reprinted, by permission, from Henry W. Hamilton and Jean Tyree Hamilton, The Sioux of the Rosebud: A History in Pictures *(Norman, 1971), pl. 45.*

Courtesy General Charles D. Roberts

Sioux Commission of 1889 at Cow Creek Agency. Left to right: John A. Lott, stenographer; Governor Charles Foster; Wilson, messenger; Major William Warner; John Warner, clerk; General George Crook; Irvine Miller, secretary; Major Cyrus S. Roberts, aide.

Congress had balked over carrying out President Hayes's promise to the Sioux chiefs, but finally, on June 20, 1878, the legislation was passed, providing for a commission to go to Dakota and settle the Red Cloud and Spotted Tail groups at agencies of their own choosing. The Sioux had waited anxiously since the melting of the snow in the spring for the appearance of this commission, and by June they were in an ugly mood, openly asserting that the chiefs had once again permitted lying white men to deceive them. The Brûlés were more stirred up than Red Cloud's Oglalas, for their tribe had split and was suffering from a division of opinion. The Corn and Loafer camp had left Spotted Tail's main camp and gone to the old Ponca Agency on the Missouri, taking with them most of the white squawmen and mixed bloods, and they were now openly preparing to settle there and take up farming. The wild Brûlés of Spotted Tail's group were infuriated over the conduct of these so-called progressive Indians. They thought it a shameful thing that Sioux warriors and hunters should be so misled by the whites that they would be willing to give up all the old customs and habits of their tribe and attempt to live like white farmers. A good sound whipping at the hands of the Brûlé soldiers was what they needed. The men in Spotted Tail's camp had revived the soldiers' lodge, thereby putting a kind of Sioux martial law in force. Even Spotted Tail had to obey the orders of the tribal council which the tribal soldiers were bound to enforce. But here lay a difficulty. The soldiers' lodge was in Spotted Tail's camp on the South Fork of White River, while the progressive Indians were at the Ponca Agency, about seventy-five miles away and out of reach of the soldiers, unless the Brûlés wished to get up a real war party and raid the camps at the agency. That would be a scandalous thing, for Brûlés to make open war on brother Brûlés. Some policy a bit more refined was obviously needed, and Spotted Tail was the man who could best think out a course of action.

The result of all this was that, in March, Spotted Tail and his chiefs visited the Ponca Agency and had a talk with Lieutenant Jesse M. Lee, who was acting agent. Lee was already convinced that the Ponca Agency was unfitted to be the final home of the Brûlés. He was easily persuaded to go with the Spotted Tail chiefs to the South Fork of White River, where he approved of the site for the new agency near the mouth of Rosebud Creek. Lieutenant Lee's action had a strong effect on Brûlé opinion, for this army officer had been in charge of the tribe since the black days of 1876, and the majority of the Indians regarded him as a staunch friend who always had their best interests in mind. The trouble was that the progressive Brûlés with their squawmen and half-blood supporters differed from the rest of the tribe, thinking only of their own interests and disregarding the opinions of both Agent Lee and the tribal council. The progressives had their minds made up, and talk was useless.

Spotted Tail now played his ace. All winter his Brûlés had refused to go near the hated agency, but in March, surprisingly, they broke camp and moved seventy-five miles to the eastward, forming a new camp on Ponca Creek, almost within sight of the Missouri and in a handsome position from which to put pressure on the Corn and Loafer camps. There were seven hundred lodges in the encampment and six thousand Indians; the camps extended for fifteen miles along Ponca Creek.[4] Spotted Tail with a small camp of personal followers had gone straight to the agency and camped there, where he could be in hourly touch with Agent Lee and keep a close watch on the Corn and Loafer progressives and the squawmen and mixed bloods. The tribal soldiers' lodge was now within close range of the Corn and Loafer camps, and even progressive Brûlés with their minds made up would probably change their views when faced by a group of grim-faced Brûlé soldiers, painted black, each man with a quirt in his right hand, a Winchester in the crook of his left arm. The quirts were for whipping disobedient Brûlés; the Winchesters to suggest what would happen to any man who resisted. Intimidation? No doubt. But the progressives and their squawmen friends were a small minority; they were trying to take control and force the tribe along the White Man's Road, whether the Brûlés wished to travel

[4] The account of the doings of Spotted Tail and his Brûlés in the spring and summer of 1878 is taken largely from contemporary Dakota and Nebraska newspapers. I am indebted to Harry Anderson for copies of notes which he took from the Niobrara, Nebraska, *Pioneer*, a little newspaper that covered these events with frequent reports.

that way or not. If Spotted Tail's retort was a bit rough, it certainly expressed the view of a majority in the tribe.

This chief was at the height of his career in 1878. Not only his own Brûlés, but most of the Sioux of other tribes looked up to him as a great leader. Among the whites his reputation was high. He had gone with a party of his chiefs and headmen to the hostile Sioux camps in the bitter winter of 1876–77 and had persuaded many hundreds of the Sioux to come to the agencies and surrender. For this great service the officials in Washington were talking of rewarding him by making him an "honorary officer" in the United States Army, with pay for life. The plan ended in talk, but Spotted Tail might have obtained that honor and the salary attached to it, as well as many other personal advantages, if he had kept close to the government Indian policy line. As it was, he held to his own belief, that you could not make good farmers out of Sioux warriors and hunters, and that the Indians would let themselves in for endless trouble and misery if they permitted the officials to have their way. Since Carl Schurz at the Interior Department had made Indian farming the first requirement in his new policy and had pledged his personal word that he would induce the Sioux to go to work on farms, it was obvious that Schurz and Spotted Tail were not going to remain good friends for many more months.

Having established his camp on Ponca Creek, Spotted Tail began to watch anxiously for the coming of the Sioux commission which was to approve the site selected for his new agency and for Red Cloud's new location. The Sioux had expected the commission to come in early spring. Congress, however, was procrastinating, half minded to discard the promise made by the President to the chiefs and order the Sioux kept on the Missouri River, by the use of military force if necessary. Week after week passed with no news from Washington. Meanwhile, it grew clearer daily that the chiefs were right and that Ponca Agency was too close to the Missouri. Quantities of clothing, blankets, and other supplies issued by the agent to the Brûlés were being traded to whiskey smugglers, and some of the Sioux camps were becoming demoralized. The settlers in northern Nebraska were accusing the

Brûlés of raiding their horse herds; the Brûlés, who were losing twice as many of their own horses, were convinced that it was the lying white settlers who were raiding their herds. The discovery that it was a gang of white horse thieves with a hideout in the Sand Hills who were stealing horses both from the Brûlés and the settlers put an end to this dispute, and Spotted Tail went to some of the towns in Nebraska with his wives and children and established really friendly relations for the time being.

In mid-June the Brûlés held a great Sun Dance on Ponca Creek, which put them in a good humor for a week. But there were rumors that the commission was not coming to move them to Rosebud, and talk was heard of their having to spend the coming winter here on the Missouri. They were growing more angry every day about this when, at last, in the first week of July, the long-awaited Sioux commission reached Ponca Agency, coming up the Missouri by steamboat. With the commission was Commissioner of Indian Affairs Hayt.

Spotted Tail seems to have taken a dislike to Hayt on sight. When a council was called, he insulted Hayt in front of all the assembled Indians and white men; he then made the threat that if his tribe was not given permission to go to Rosebud within ten days he would burn Ponca Agency and go to Rosebud. After the public council he met the commissioners in private and told them that he had spoken the will of the tribe but that personally he had only the most friendly feelings toward the gentlemen. Still, his people were very much excited, and if the move to Rosebud were not ordered soon, there might be serious trouble. The chief seems to have felt that this very important matter of establishing his people at a new agency was being bungled, as nearly all government actions had been bungled in the past. His Brûlés had been on the reservation since 1868, and in those ten years the agency had been moved four times. Each removal had cost the government a great deal of money. Each new agency site had been approved by inspectors or a special commission and had been found later to be most unsuitable. There was every reason for the present commission to examine the new site at Rosebud carefully before making a decision. But Spotted Tail's Brûlés were in a hurry to move; the commission was very late in coming and had to press forward to meet Red Cloud and his chiefs, and so it accepted the Rosebud location as a fine site for the Brûlés simply because Spotted Tail and his acting agent said that this was so.

With its habitual ineptitude, the Indian Office chose this moment of crisis for changing Spotted Tail's agent. With a great need for an experienced man in charge who knew the Brûlés well, it gave an appointment as agent to James Lawrence, apparently for political reasons alone. Lawrence had been agent for the Poncas and had been involved in the scandals that resulted from the forced removal of that tribe to Indian Territory. For some reason of his own, Lawrence refused the appointment as Spotted Tail's agent, and at the last moment W. J. Pollock was appointed as acting agent and hastened to Ponca Agency to take charge. Pollock was a Western man. He knew Indians well and was personally known among the Brûlés, but he found a serious situation at Ponca, and Commissioner Hayt was not making matters any easier by assuming an attitude of opposition to the removal to Rosebud. Ordered to delay, Pollock appealed to Spotted Tail, who advised him to employ the Fox Soldiers. This was a very strong Brûlé brotherhood with a membership of three hundred warriors. Pollock hired them by giving them free government beef, and they policed the Brûlé camp, doing all they could to prevent any groups starting for Rosebud. But the Indians were determined to leave the hated Missouri, and on July 22, Pollock wired the Indian Office that part of them had left Ponca and the rest were leaving the next day. On the twenty-ninth, he wired that all Indians had left and that troops were needed to follow them. Secretary Schurz would have preferred to resign, however, rather than to call for troops. His whole policy was based on an effort to show the country that he could handle the Indians without the aid of the military.

Commissioner Hayt sent a scolding telegram to Pollock. He demanded the reason why Pollock had not persuaded his Indians to wait at Ponca. Pollock wired back angrily that he was not dealing with white people who were open to reason but with wild Indians, and that the time for persuading them to remain had passed before he was appointed agent. The Indians had brushed aside his advice and were now encamped one-third of the distance to Rosebud. A large group of the old, ill, and crippled was trailing along behind the main body, on foot, cursing the officials who had forced them to come to the Missouri and who now failed to provide wagons to transport them to the new agency.

Reports came back to Ponca that as the Brûlés moved toward Rosebud they were quarreling violently. Apparently the Corn, Loafer, and squawman groups (who wished to remain near the Missouri and farm) were resisting being moved to Rosebud, and Spotted Tail and his tribal soldiers were forcing them along. There had been actual fighting; some Sioux had been shot; and eight hundred young warriors were reported to have ridden off northward, on what mission no one could guess. Part of the Sioux seem to have run away at this moment, for they later turned up in Sitting Bull's hostile camp in Canada.[5] Hayt had cancelled a contract for wagon transport on the grounds that the rate demanded was exorbitant, forgetting that at this moment any wagon freighter who took a contract would have to risk having his men shot as a part of the bargain. Pollock had got together enough teams and wagons to send some rations and other supplies after the moving Sioux camps; he then took the trail. Reaching Rosebud August 29, he reported the Indian camps were one day's march behind him.

During these tumultuous days Spotted Tail proved again his powers of leadership. He was the one man who held the Brûlé tribe together, preventing it splitting into angry factions and going off in several directions. Part wished to stay at Ponca, part to remove to Rosebud; others were determined to run away and join Sitting Bull in Canada. They quarreled day and night about everything. The White Robes (Protestant Episcopal missionaries) and the Black Robes (Roman Catholics) came into it. Spotted Tail and his chiefs were annoyed with the Episcopalians, who had had a missionary at the old agency as far back as 1875. The chiefs said the White Robes had not taught one Sioux child

[5] These were northern Indians, hostiles who surrendered in 1877. Agent Newell reported their return to Rosebud in February, 1880.

to speak or write English. As for the teaching of Christianity, that did not interest most of the chiefs. All they desired was to have some full-blood Sioux boys taught to read and write English, so that they might act as interpreters and also write letters from the chiefs to the officials in Washington. When the march to Rosebud started, the chiefs refused to permit the Episcopal missionary to go with the Indians; but two Roman Catholic priests in some manner slipped into the Sioux camp and went along. When Agent Pollock reached Rosebud, the chiefs held a council with him and stated in strong terms that they wished to have the White Robes barred from the new agency, but wished to have the Black Robes start a school.[6]

The Sioux commission that was supposed to inspect the new agency site at Rosebud had gone up the Missouri by steamboat. They landed at the new but already useless Red Cloud Agency, on the west bank at the mouth of Medicine Creek. The site of this agency had been carefully selected by an official group and the agency had been built at great expense. The new commission made an inspection and pronounced the site a poor one, indeed, an impossible location if Red Cloud's Oglalas were expected to make a living by farming.

Going up White River, the commission came to Red Cloud's camps. Here there were no displays of anger and discourtesy such as had greeted the commission at Spotted Tail's Ponca Agency. The Oglalas were not split as the Brûlés were. They had united in selecting a site for their new agency on Big White Clay Creek in the edge of the Pine Ridge, and Red Cloud met the commission with his best company face, smiling and genial. He and a group of headmen escorted the commission up White River to the site of the new agency, but after inspecting the location, the commission began to have serious doubts of the wisdom of placing a great Sioux agency in such a spot. The site was as far from the hated Missouri as the Oglalas could go without leaving their res-

[6] I am indebted to Harry Anderson for copious notes on the removal from Ponca to Rosebud, which he obtained from the National Archives and from the Niobrara, Nebraska, *Pioneer*. My thanks are also due to E. W. Sterling of the University of South Dakota for notes on these events from the Yankton papers of the day.

ervation. That had evidently been the main reason for their preference for the spot. With no intention whatever of trying to farm, they had chosen a fine location for Indians to camp in. The commission pronounced the land too sandy for farming; they preferred Wounded Knee Creek, to the east and in the center of the future Pine Ridge Reservation, and they left an urgent request with the military officer who was to superintend the building operations that the agency should be at Wounded Knee. The usual official muddling ensued, the military officer regarding the site of the agency to be a matter outside his field of duty, and the Indian Office officials forgetting all about it. Meanwhile, Red Cloud was on the spot with his chiefs, repeating daily that they would have their agency here at Pine Ridge, not at Wounded Knee, and that they would make abundant trouble if their wishes were ignored. By sheer weight of reiteration they won their point, and the agency was built where they wanted it.

Traveling eastward, the commission inspected the site of Spotted Tail's new agency at Rosebud on the South Fork of White River. This stream had a good flow of clear water running over a gravelly bed; there was fine timber along the stream and much good farmland in the valley; but when the commission saw the site on Rosebud Creek that Spotted Tail and his chiefs had selected for the new agency, they were aghast. It was a beautiful spot for a Sioux camp, on fine grounds overlooking Rosebud Creek, and surrounded by picturesque hills handsomely dotted with small pines—but it was a very bad location for a great Indian agency. The site lay high above the creek, while at the same time it was down in the bottom of a deep bowl rimmed by hills. There was no water on the agency site and no way of hauling in rations and other supplies by wagon except over the sandy and steep hillsides, which in bad weather would be almost impossible with heavy loads. Yet the Brûlés would not consent to having their agency at any other point. The exasperated commissioners did their best to talk the chiefs around; then gave it up and departed.[7]

Thus the Red Cloud and Spotted Tail Sioux were settled in

[7] The report of the Sioux commission of 1878 is in the *Annual Report of the Commissioner of Indian Affairs* (1878), 160.

their final locations. The story of the movements of these two Sioux groups on the reservation was a shocking exhibition of government inefficiency. Spotted Tail's agency had been moved five times, Red Cloud's three times, each move costing very large sums. From the first, the government purpose had been to locate these Indians on land where they could take up farming and where they could be supplied without prohibitive costs for wagon freighting. In every instance, partly from official bungling, partly from Sioux stubbornness, the Indians had gotten what they desired—a nice place for them to camp in, where the land was hardly fit for farming and was so far from a base of supplies that the wagon transportation about doubled the cost of every pound of supplies sent to the agencies.

The moment the new agencies were established in 1878 this matter of the high cost of wagon freight came up again. Spotted Tail's agency at Rosebud was about one hundred miles (by wagon) west of the steamboat landing on the Missouri, which was to be the new base of supply, and Red Cloud's agency at Pine Ridge, west of Rosebud, was much farther away. In July, Commissioner of Indian Affairs Hayt called for bids for wagon freighting from the steamboat landing to the agencies. The Dakota men had a monopoly of this business and thought that they had the Indian Office at their mercy. When the bids were opened, they proved to be so high that Hayt was at first horrified and then angry. The wagon freight from the landing to the agencies was going to cost more than the original price of the rations and other goods, including their transportation to Rosebud Landing on the Missouri. Commissioner Hayt denounced the Dakota men and stated that he would not permit them to rob the government in this manner. But how was he to get out of this dilemma the stubbornness of the Sioux had placed him in?

Someone in the Indian service remembered that at this moment a very large government-owned train of big freight wagons with ox teams was at the old Red Cloud and Spotted Tail agencies near the head of White River, idle and useless. Why not bring the wagon train to the Missouri, turn the wagons over to the Sioux, and have the Indians haul their own rations and supplies

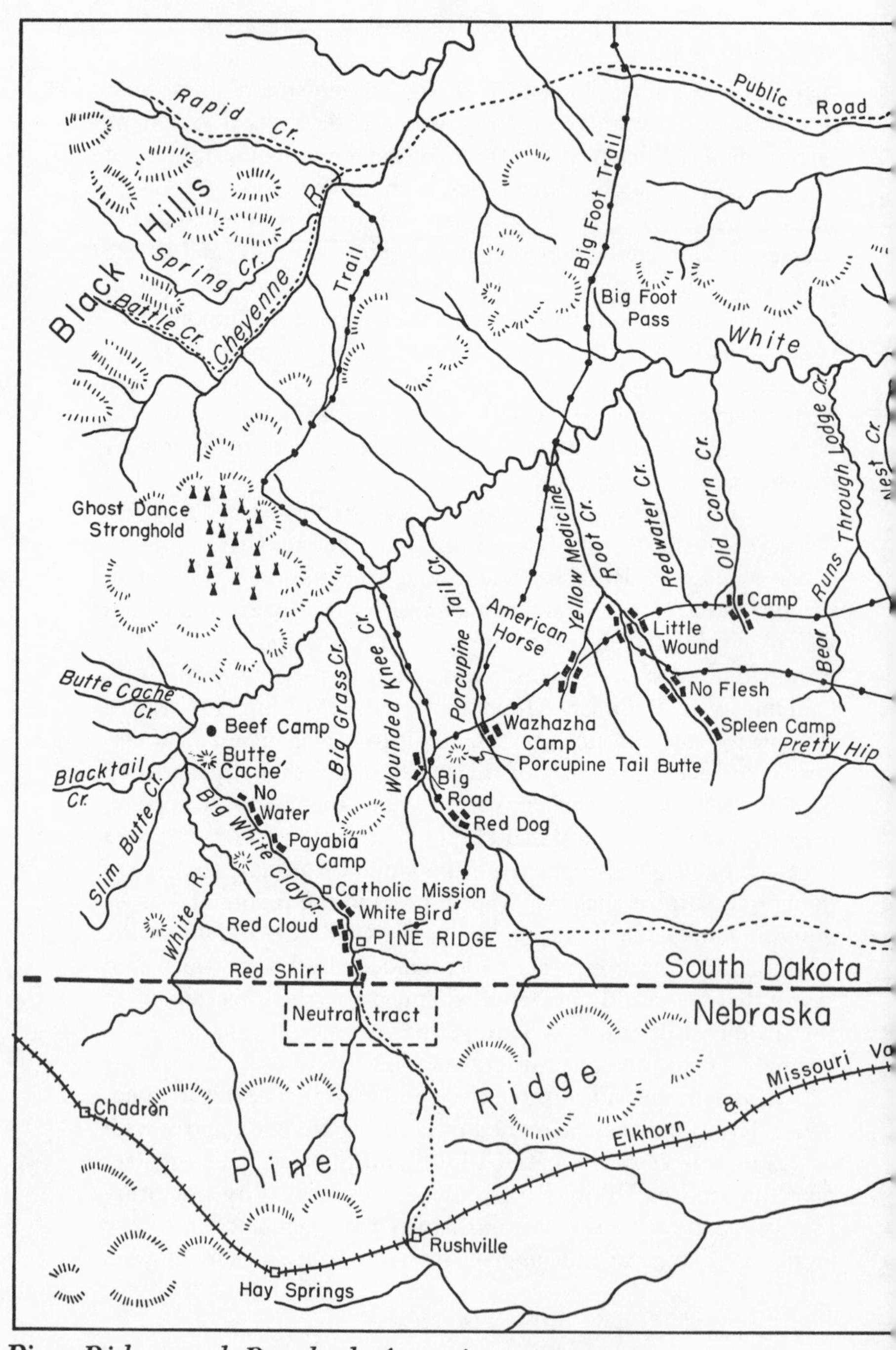

Pine Ridge and Rosebud Agencies, 1878-1890

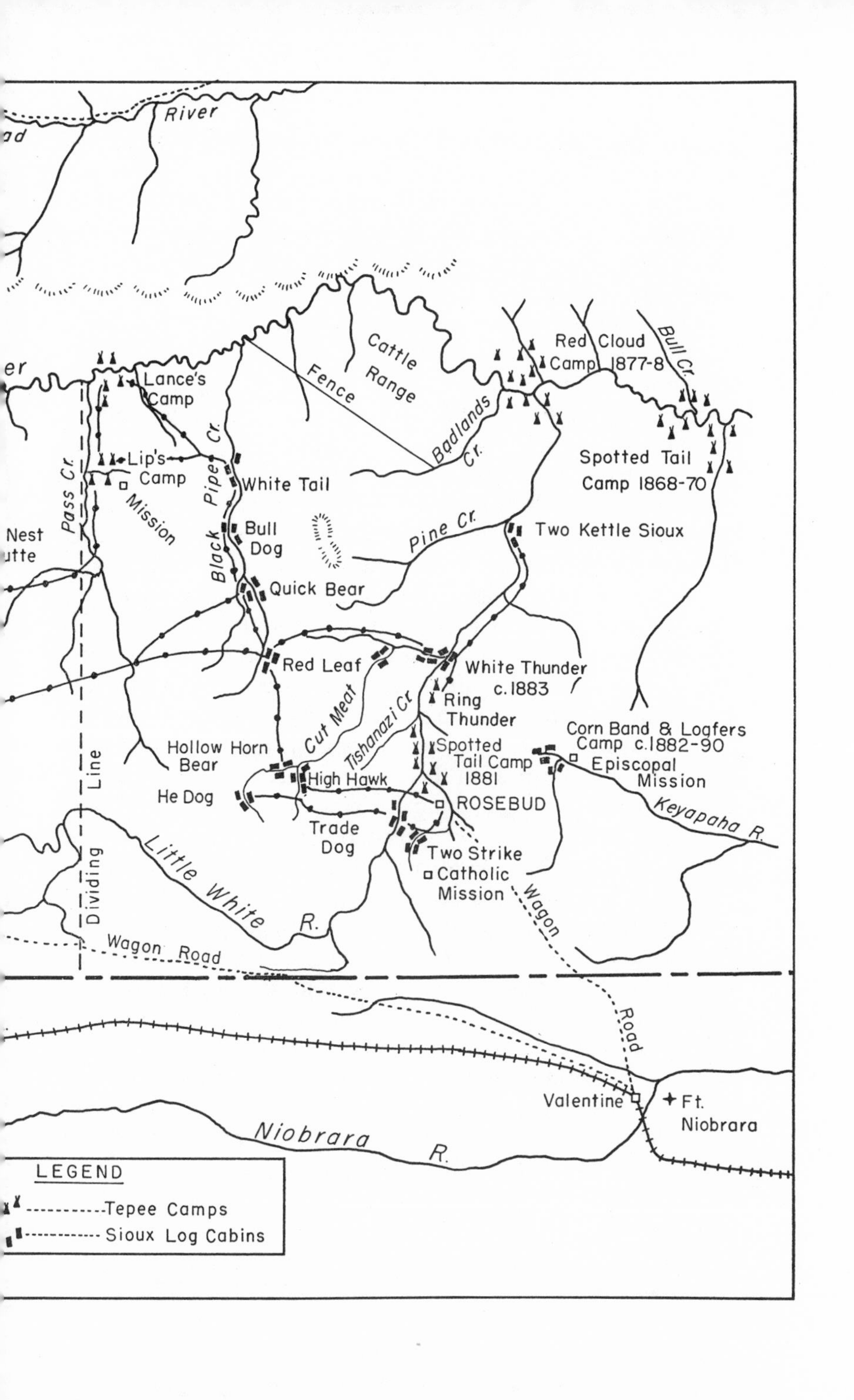

River
Cattle Range
Fence
Red Cloud Camp 1877-8
Bull Cr.
Lance's Camp
Badlands Cr.
Spotted Tail Camp 1868-70
Lip's Camp
White Tail
Mission
Pass Cr.
Black Pipe Cr.
Pine Cr.
Bull Dog
Two Kettle Sioux
Quick Bear
Red Leaf
White Thunder c.1883
Cut Meat
Tishanazi Cr.
Ring Thunder
Corn Band & Logfers Camp c.1882-90
Hollow Horn Bear
Spotted Tail Camp 1881
Episcopal Mission
High Hawk
He Dog
ROSEBUD
Keyapaha R.
Trade Dog
Two Strike
Catholic Mission
Dividing Line
Little White R.
Wagon Road
Wagon Road
Valentine
Ft. Niobrara
Niobrara R.
LEGEND
Tepee Camps
Sioux Log Cabins

from the steamboat landing to the new agencies? All the officials were anxious for these Indians to go to work, and here was an excellent opportunity for making a start. Hayt promptly adopted this plan, and within a few weeks wagon trains of supplies, manned by Sioux Indians, were on their way from the Missouri River steamboat landings to the two new agencies. The government supplied the big freight wagons and teams. The Indian ponies were too light in weight for such work, and the Sioux men did not know how to harness or drive the teams; but, with aid and instruction from some of the white squawmen and mixed bloods, the fullbloods soon learned. They liked the work. It meant traveling, a thing dear to their hearts; they took their wives and children on the wagons, and at night the Sioux teamsters and their families camped in tipis at the trailside and gathered at their campfires. To add to their content, at the end of each trip every driver received about twenty silver dollars in pay. Commissioner Hayt made a great to-do over what he termed this miraculously sudden advance of the Sioux toward self-support, and he and the other officials got all the credit. In plain truth, Spotted Tail, eight years back, had dictated a letter to be written by the trader Joseph Bissonnett, in which he urged on the officials this identical scheme of having the Indians haul their own supplies to the agency, for the double purpose of teaching the Sioux how to work and saving the government money on wagon freighting. The officials had pigeonholed the chief's letter and gone on for eight years, paying exorbitant wagon-freight rates to white contractors.[8]

Four hundred wagons were soon employed in the work, and during the autumn of 1878 the Spotted Tail and Red Cloud Sioux hauled four million pounds of freight from Rosebud Landing (the place was called Black Pole by the Sioux) to their agencies, receiving one dollar per hundred pounds per hundred miles for the work. They earned altogether about $120,000. Best of all, they exploded the old belief that none of the Sioux men would work with their hands. True, the Sioux engaged were almost all from the more progressive camps, and this work was made to order to

[8] Spotted Tail letter, quoted by William Welsh in his pamphlet on his visit to the Sioux in 1870.

fit exactly the Indian idea of congenial labor. Except for the loading and unloading of the wagon, in which the white employees helped, there was no work, and during the three-day haul from the Missouri to Rosebud (a day or so farther to Pine Ridge) the Sioux, with their families riding on the wagons, drove happily across the trackless prairie, camping where they chose to at night. Most of the drivers made a trip or two; then (rich with silver dollars) they quit, handing over their job to some brother or cousin; thus about eight hundred Sioux were presently engaged, from time to time, in freighting goods from the landings on the Missouri. The work was kept in certain families to a large extent, and with excellent results, for the Sioux learned by experience, and—being urged to use their own teams—they were soon cutting hay and even grew some corn, and their Indian ponies were astonished to find themselves for the first time in history sheltered in log stables and properly fed, instead of being turned loose on the prairie to shift for themselves, summer and winter.

Thus in 1878 the last of the Western Sioux were settled in final locations—Spotted Tail's Brûlés at Rosebud, Red Cloud's Oglalas to the west of them at Pine Ridge. The other agencies were Lower Brûlé, Crow Creek, Cheyenne River, and Standing Rock. Lower Brûlé (the Sioux here were close kinsmen of Spotted Tail's Upper Brûlés) was on the west bank of the Missouri, north of White River. On the east bank of the Missouri, opposite to Lower Brûlé, was Crow Creek Agency, where the Lower Yanktonais Sioux were located. Farther up the river, on the west bank near the mouth of Cheyenne River, was Cheyenne River Agency, for the Two Kettle Sioux, the Blackfoot Sioux, the Miniconjous, and the Sans Arcs; while farther north, on the west bank and extending to the Cannonball River in North Dakota, was Standing Rock Agency, for the Upper Yanktonais, Hunkpapas, and part of the Blackfoot Sioux. The only Western Sioux who were not now at these six agencies were in Sitting Bull's camp, which had fled across the Canadian border during the Sioux war in 1876. These Sioux were pretending valiantly to be the only free Sioux left in the world, although the redcoats of the Canadian Northwest Mounted Police were in fact controlling their actions more strictly than the bluecoat soldiers of the Great Father ever had attempted to do.

The commission that had been sent out to settle the Spotted Tail and Red Cloud Indians in locations of their own choosing was far from satisfied with the selections made by the Sioux, for it did not regard the sites at Rosebud and Pine Ridge as suitable ones from the point of view of farming. However, the locations were as good as the sites the government officials had selected for the other agencies along the Missouri. In plain truth, there was little land on the great Sioux reservation that, in its wild state in 1878, any Eastern farmer would have regarded as really good farm land. Added to this was the fact that the entire region was so subject to savage droughts and to invasions of countless hordes of grasshoppers that even experienced white farmers could not expect to average more than one good crop in every three or four years. Some of the Sioux who lived along the Missouri had been trying to grow crops for many years, but they had never achieved anything beyond an occasional small supply of corn and vegetables, to be added for a brief period to their regular diet of meat. Now that buffalo and other big game had disappeared from the vicinity of the Sioux reservation, all these Indians were depending on free government rations.

The idea that the Sioux could suddenly jump ahead into complete self-support from farming had been concocted by theorists and humanitarians in the Eastern states: men and women who were meddling in Indian welfare and who just could not comprehend that Indians were not like white people. Tribes like the Sioux had been hunters and warriors for thousands of years; they knew no other manner of making a living, and they had it grained into their flesh and bones that people who settled down on one spot and tried to live by growing crops were weak and contemptible. Thus most of Spotted Tail's Brûlés and Red Cloud's Oglalas looked down on such groups as the Brûlé Corn Band and the Sioux at the Missouri River agencies, who were trying their hands at growing crops. They regarded these tame Sioux as weaklings and traitors who had deserted their people and the old way of life and were trying to imitate the ways of white people. Thus the few

Sioux who in 1878 were making their faltering attempts at farming, most families with a farm of less than one acre, not only had to fight against the Dakota droughts and grasshopper plagues but against the solid opposition of their own people. Thus when Carl Schurz in Washington pledged his word to Congress that the Sioux would very soon be self-supporting through farming, he was talking nonsense.

2

Spotted Tail Departs

AT PINE RIDGE (called *Wazi Ahanhan* by the Sioux) Red Cloud and his Oglalas were happy through the year 1878 and on into the winter. The difficult days of 1876–77, when Red Cloud and his folk had been under the military supervision of General Sheridan and his troops, were gone and largely forgotten, and the prospects were now such that Red Cloud could look forward to prosperous times, with his declining years brightened by the high regard in which he was held among the Sioux and among the white people in the Eastern states. Having selected the site for their new agency on Big White Clay Creek—as far as they could get it from the hated Missouri River without placing it outside the Sioux reservation and in Nebraska—Red Cloud and his followers sat in their tipis, watching the white men put up the buildings for the agency, visioning long years of peace and good living ahead. Then, in March, 1879, Red Cloud was struck one of the shrewdest blows that blind fate ever aimed at a Sioux chief. Dr. Valentine T. McGillycuddy was appointed to be his new agent.

McGillycuddy, then a surgeon in the army with the rank of major, had been stationed at Camp Robinson, Nebraska, within a mile of the old Red Cloud Agency in 1876–77, and there he and Red Cloud had met and taken a notable dislike to each other. Red Cloud had an aversion toward army officers, and he held in particular detestation the type of stiff and haughty military martinet, of which class Major McGillycuddy was an outstanding example. Still under thirty, tall, broad-shouldered, and vigorous, McGillycuddy was well educated but surprisingly narrow and bigoted in his views (particularly in those that concerned the Sioux Indians). He was a firm believer in Prussian discipline and was a stiff and haughty upholder of the faith that officers and gentlemen were a class apart. To Red Cloud all this was anathema. He regarded McGillycuddy as an obnoxious boy, and now this boy was to be his official father, his new agent!

The Sioux called McGillycuddy *Putin hi chikala*—Little Whiskers—a name which really referred to his mustache. He was clean-shaven except for that; but the mustache was an outstanding feature of his handsome face. It was the type favored by cavalry officers of the day, a great inverted "V" the ends of which reached two inches below the corners of his mouth. It was a badge of rank, rating the wearer as a commander and a fighter. McGillycuddy easily qualified for both these ratings.

When told the news, Red Cloud at first could not credit it. He had met the new commissioner of Indian affairs in August, and his relations with Hayt had been cordial. Why should this friendly bald-headed man strike such a blow at him? The chief did not know that it had been Carl Schurz who had appointed McGillycuddy and that the whole thing was just a piece of sheer bad luck. As a liberal, Schurz disliked the class of aloof and haughty regular army martinets as much as Red Cloud did; but now, as secretary of the interior, he had many official contacts with the military. In 1878 the Cheyennes tried to fight their way home from Indian Territory to Montana, and Schurz had to confer frequently with army officers. Knowing almost nothing about Indians, he permitted the military officers to convince him that if the Red Cloud Sioux did not have a strong agent, preferably an army officer, in control of them, they might be drawn into the Cheyenne hostilities, and thus a new Indian war in the plains would come about. This view of the situation was absurd; but once it was implanted in the mind of Schurz, it stuck. Then by pure accident he met young Major McGillycuddy, fresh from the scenes of fighting with the Cheyennes near Camp Robinson, Nebraska. The impressionable Secretary of the Interior listened enthralled to this officer's vivid descriptions of the struggle with the Cheyennes, and—taking a sudden liking to the handsome and bold officer—Schurz surrendered to his penchant for making quick decisions and offered McGillycuddy the post as agent for Red Cloud's Sioux. On such occasions as this Carl Schurz acted on the impulse of the moment, leaving innocent bystanders to pay for it later on. In this particular case, Red Cloud and his Oglalas were the innocent bystanders, and they paid through seven long and bitter years.[1]

While those two leading liberals, Carl Schurz and Ezra Hayt, were encouraging Congress to provide funds necessary for a farming crusade among the Sioux, the Oglalas and Brûlés were sitting in their tipi camps, contentedly eating free rations. Out of a population of perhaps fifteen thousand, some eight hundred young men were doing a little work in hauling supplies from the Missouri River and the Union Pacific Railway. No one else had lifted a finger, except for a few of the white squawmen and a little group of Brûlés of the Corn and Loafer bands, who were trying their hands at gardening, not farming. Most of the families were planting a fraction of an acre. In his first enthusiasm Hayt had spent a large sum in having forty-acre tracts surveyed and marked for Indian farming at Rosebud and Pine Ridge; but these little farms all lay untouched, and when a special commissioner was sent out to urge the Sioux to leave their tipi camps at the agencies and move to the farming tracts, he had only fair success at Pine Ridge, and at Rosebud none at all. And this was the advance in civilization and self-support which the Washington officials were describing as simply marvelous.

The truth was that both Schurz and Hayt had been strongly influenced in their view concerning Sioux prospects by the excited talk of the Christian and humanitarian leaders who at this period were exerting steady pressure on the officials in Washington and on Congress for a liberal Indian policy and lavish spending. These men were filled with blind faith that, if given some tools and a few acres, every Sioux head of a family would leap

[1] Mrs. McGillycuddy in her book, *McGillycuddy Agent*, gives a stirring account of her husband's appointment as agent by Schurz. But it was not true that Red Cloud was siding with the hostile Cheyennes, and army men knew it. Red Cloud sent his son-in-law American Horse to the scene of fighting, not to aid the Cheyennes but to offer any assistance he could give to the military.

suddenly into self-support. They had sold these notions in Washington in 1868, and the experiment had proved a terribly costly failure. Now they were back again, eleven years later, exerting such pressure that no one in official circles could touch on Indian affairs safely unless he spoke the language and shared in the faith of these crusaders. Carl Schurz accepted the views and the policy put forward by these men and made himself personally responsible to Congress for a successful issue.

On the Sioux reservation the Indians exhibited no such yearning toward progress in civilization and toward self-support as these men in the East attributed to them. In 1879 the matter that really interested the Sioux and their agents was *control.* Congress on May 27, 1878, had passed a seemingly unimportant bill providing for the recruiting of Indian police forces on all reservations, the police to be under the orders of the agents. The Sioux chiefs were illiterate men, but they were no fools; although they did not comprehend at once the government's purpose in raising forces of uniformed police, they soon began to suspect the truth —that the Indian police were to be employed in breaking the chiefs and placing control entirely in the hands of the agents. That was exactly the idea back of the program, which had been thought out by the self-styled Indian Friends, who by 1877 had given up their earlier fervid view that Red Cloud, Spotted Tail, and the other chiefs were heroic leaders of their people in favor of a new and equally fervid belief that these chiefs were selfish and reactionary despots who were determined in their own interests to prevent any change in their people's way of life. In effect, the chiefs would not co-operate with the white crusaders in the effort to hustle the Sioux along the way they had labeled "progress," and with their usual lack of sincerity the Indian Friends pretended that in this matter they were taking the side of the common Indians against the despotic chiefs. They drew a dark picture of Indian tribal government that was a travesty on truth. At this moment the common men among the Sioux were very much perturbed over the new conditions they had to face on the reservation, and they were anxious that their chiefs should be left free to advise and lead them. The alleged despotism of the chiefs was a myth. In any crisis the tribal council made its decisions, and Red Cloud, Spotted Tail, and the other chiefs were ordered to give utterance to those decisions. In an acute situation the Sioux put the soldiers' lodge in control and the chiefs for the moment were put back into the bag (as the Indians expressed it). This was like martial law, with the council and soldiers in absolute control. In the name of democracy the white crusaders were proposing to destroy a fairly free form of Indian government and to replace it by a paternal despotism with white officials in complete control. This they were going to do for the good of the Indians, or so they imagined.

The effort to start this new system of control over the Sioux had hardly gotten under way in 1878–79 when it became apparent that the Indians were so bitterly opposed to the whole program that only agents who were willing and able to ride roughshod over their Sioux could hope to organize efficient Indian police forces. At Cheyenne River Agency, Captain Theodore Schwan was the agent. He was a German. His family had come to America when he was a small boy; he had enlisted as a private in the old regular army in 1858; and during the Civil War he had won the Congressional Medal for heroism and had been promoted to the rank of captain. He believed in discipline, and when he was appointed agent for the Sioux at Cheyenne River, he took a firm grip on things.

The Sioux at his agency had been roughly treated in 1876 when, as a part of General Sheridan's war measures against the hostile Sioux, the friendly, or at least peaceful, Indians at this agency had been disarmed and their ponies taken from them by a large force of cavalry. They had been kept under strict military control until Captain Schwan was made agent, and the Sioux found that Schwan was a sterner master than the army had been. They called him The-Man-Who-Never-Smiles. He was the first Sioux agent to recruit a force of uniformed and armed Indian police, and after that the only way in which his Sioux could have any freedom or ease was by camping many miles away from the agency and only coming there once in ten days to collect the free rations that were their due under the treaties.

When this matter of the organization of Indian police forces came up in 1878–79, it became apparent that most of the Sioux close to the Missouri were more easily controlled than were the freer bands under Spotted Tail and Red Cloud. This was partly due to the fact that most of the Missouri River bands were more familiar with the whites and their ways than were the wild Brûlés and Oglalas, and that during the war of 1876 they had been dismounted, disarmed, and kept under military control. Even in 1879 there was an army post, Fort Bennett, within sight of Cheyenne River Agency, and just across the Missouri was the larger post, Fort Sully. Up the river at Standing Rock the post of Fort Yates was firmly established at the agency. Below Cheyenne River were Fort Hale and Fort Randall.

Thus when the Indian Office ordered the forming of police forces at the agencies, the Sioux at Cheyenne River submitted, as did also those at Standing Rock. Down the river at the little Crow Creek Agency on the east bank, the tame Yanktonais Sioux deserted their chiefs (for the time being) when their military agent, Captain William G. Dougherty, ordered the recruiting of a police force. Dougherty had two agencies, Crow Creek on the east bank and Lower Brûlé on the west bank. The Lower Brûlés were the parent group from which Spotted Tail's Upper Brûlés had split off and wandered westward as far as Wyoming and south into Kansas. Compared to the Upper Brûlés, the Lower Brûlés were a rather tame lot, but they had a reputation of being the wildest of the Missouri River Sioux. When Agent Dougherty attempted to organize a police force at Lower Brûlé, he ran into stubborn opposition. Persisting, he did get a small force of uniformed men together; but there was something wrong and the police were ineffective. At Rosebud, Cicero Newell, a weak agent, accomplished nothing. At Pine Ridge, Agent McGillycuddy after a violent effort succeeded in organizing a strong police force, and thus was started his seven years' war with Red Cloud.

From the moment he took charge at Pine Ridge in 1879, McGillycuddy was a marked man. His energy and ability were apparent from the first, but he had other striking qualities that soon won for him a host of friends and an equal number of enemies. He was, unlike most of the Sioux agents, a gentleman, cultured and steeped in the regular army cult for those little formalities and punctilios which one gentleman rendered willingly to another and expected to receive as his due from inferiors. This agent would not brook the slightest familiarity of conduct either from Indians or from the whites at the agency; if any of these people dared to disobey him, he would not rest until he had taught them their lesson or driven them from the reservation. He greatly improved the agent's residence at Pine Ridge, and Mrs. McGillycuddy (his first wife) turned it into a home in which cultivated people could dwell in comfort. His enemies hinted that he was using Indian funds to provide himself with a palace. McGillycuddy met the charge with explosions of invective and broadsides of biting sarcasm. The agent stated, among other things, that he considered it his due that he should be permitted at least to live like a gentleman. The other side retorted that the "layout" at Pine Ridge was more like that of a feudal baron.

In both speaking and writing, McGillycuddy had a most effective style: terse, biting, and at times highly humorous. When engaged in one of his innumerable feuds, he simply diffused haughtiness, scorn, and ridicule in every direction, assailing his opponents with a vigor and variety of language that generally left them speechless and aghast. Even in his official reports he displayed an originality in phrase and treatment that made his writing stand out in marked contrast with the dull and plodding compositions of the other Sioux agents. In his very first report we find this:

Through carelessness or design, and directly against the orders of the Interior Department, this agency was, in the fall of 1878, located in the southwest corner of Dakota, within 1¾ miles of the Nebraska line, so that when I assumed charge here in 1879 we were furnished with the luxuries and accommodations of civilization by having a well supplied whisky ranch in full blast, almost within gunshot of the agency, which forced the agent to add the labor of coroner and undertaker to his other duties by making periodical trips into Nebraska to gather up dead Indians and half-breeds, killed in drunken quarrels.

This single sentence produced an uproar, certain Indian Office officials considered that the first four words aspersed their honor, while wrathful Nebraska frontier editors commented bitterly on this agent's false statements concerning the activities of the citizenry on the Pine Ridge border. McGillycuddy did not care. He adored a fight.

Before he had been at Pine Ridge for a month, he had brought up the burning question of an Indian police force. The Sioux resisted; McGillycuddy resorted to pressure; the Sioux continued to resist, and the pressure was increased. In the end he enlisted the largest police force at any Sioux agency, over the opposition of practically the whole Oglala tribe. The feeling of friendliness vanished from Pine Ridge, and the Sioux watched sullenly as the agent paraded his new armed force. Led by Red Cloud, sixty chiefs and headmen signed their marks to a petition, asking the Secretary of the Interior to remove the agent. The *New York Tribune* denounced McGillycuddy as a tyrant; the Quakers of Philadelphia had the same idea; but the agent had plenty of partisans, who denounced his enemies as silly sentimentalists (this was aimed at the Quakers), crooked "Indian Ringers," and Indian reactionaries (meaning Red Cloud and his chiefs). These friends insisted that McGillycuddy was the best agent among the Sioux; his agency was in fine condition; and the progress since his coming was remarkable.[2] One finds it difficult, however, to discover just where this progress lay. The agent had recruited a police force and established iron discipline among the Oglalas; but this gain, if it was gain, had been won at the cost of good feeling and friendliness. Some of the bands had left their tipi camps at the agency and moved to stream valleys, where a few of the families were building rude log cabins, but this accomplishment was mainly the work of a special agent, sent from Washington to coax the Sioux into scattering out on farm lands. There was a tiny day school at the agency, and some more were to be established in other parts of Pine Ridge, but pressure had to be brought on most of the Sioux before they would put their children in school, and the

[2] *McGillycuddy Agent*, 51–53. Here again Mrs. McGillycuddy tells the story entirely from her husband's point of view.

police were very active in enforcing attendance. That was all the progress. Other Sioux agents were doing as much, or more, and without filling the nation's newspapers with angry debate.

Red Cloud, who did not like white soldiers, was particularly exasperated by the military parade that McGillycuddy had inaugurated at his agency. The new Indian police were uniformed and drilled like soldiers; they mounted guard with ceremony; there was a guardhouse, a tall flagpole and a flag, and the whole agency bristled with painted notice boards and tacked-up rules. Like an officer in command of a garrison, McGillycuddy was regulating every little activity of both the white employees and the Sioux, and new sets of rules and regulations were constantly being put into force. Visitors on the reservation had hardly crossed the line from Nebraska before they were met by warning notices to report themselves immediately at the agent's office, give an account of themselves, and deposit any weapons they might have in their possession. During the whole of their stay they were spied upon by the agent's police, and if they broke the smallest of McGillycuddy's rules they were instantly called to account.

About a year after McGillycuddy's coming to Pine Ridge, the ranchmen in Nebraska and eastern Wyoming began to suspect that some men at Pine Ridge were stealing horses, and Edgar Bronson (a member of the cattlemen's association) suggested to the new agent that an expert brand reader should be sent to the agency to keep a watch. McGillycuddy approved the idea, and Bronson sent a veteran cowboy to Pine Ridge—a man of character and cool courage, a reliable man of the type of Owen Wister's Virginian. A short time after this Bronson was amazed to receive a curt note from McGillycuddy, demanding that the brand inspector be taken away and threatening that if this was not done at once, he would have his Indian police remove the man from the reservation. At the same time an equally amazing note came from the brand inspector, stating that he did not wish to remain at Pine Ridge any longer since he was having difficulty in restraining himself from shooting the agent. Hastening to the agency, Bronson made discreet inquiry and discovered that this threatened outbreak of war had been brought on by the brand inspector (who liked and admired McGillycuddy) going into the agent's office uninvited, sitting down without being asked to do so, rolling cigarettes and lighting them without permission, and, on occasion, tilting his chair back and resting the high heels of his spurred boots on the edge of McGillycuddy's desk. Not a word had been said by either party, but the haughty agent had simply filled the air of the office with frozen particles until the brand inspector, noting the fall in temperature and seeking its cause, had reached the conclusion that for some unknown reason the stiff individual behind the desk did not like him. He had soon reached the further conclusion that he did not like the agent and had better shoot him. Bronson thought this all highly amusing; but Red Cloud and the white men who had to live at Pine Ridge and face McGillycuddy's haughty looks and sudden acts of temper, could see no humor in it. This agent was in absolute control of their fates, and (like the brand inspector) many of them were with difficulty restraining themselves from shooting him.

Yet, for outsiders, the situation at Pine Ridge was often funny. McGillycuddy, on orders from Washington that neatly fitted in with his own desires, was trying to destroy the power of the chiefs, break up the tribal organization, and gather all authority into his own hands as United States agent. But he could not even talk to his Indians unless he requested the chiefs to call a council, and even then he had to talk through interpreters who were generally chosen by the chiefs. In frequent public utterances, McGillycuddy denounced the chiefs as tyrants who kept the common Indians in subjection; yet at every council these common Indians placed themselves behind the leaders of their own blood, in open opposition to this white man who was trying to save them. The agent lost the first rounds in the police struggle when the Sioux almost to a man sided with their chiefs in opposing a police force, and it was only after weeks of scheming, urging, quarreling, and the making of threats that McGillycuddy finally prevailed and recruited a police force. At the time McGillycuddy did not dwell on the exact methods by which he triumphed over Red Cloud and the other chiefs, but he implied that the common Indians—anxious to be protected by the police from the tyranny of their chiefs—finally threw their support to his side. In later years he was more frank, admitting that from the first all of the Sioux, chiefs and commoners, were bitterly opposed.

It was soon apparent that there would be no peace at Pine Ridge as long as Red Cloud and McGillycuddy were there, each striving for mastery. At frequent intervals the newspapers were filled with reports of the uproarious doings at this agency, and solemn predictions were made that this agent, or Red Cloud, if not stopped would soon bring on open war. The two contentious gentlemen seemed to be too angry to care what happened; but whenever a crisis was reached, each of them had sufficient sense not to take the final step that meant real battle.

While Red Cloud was thus sorely beset at Pine Ridge, at the neighboring Rosebud Agency his old rival Spotted Tail was being made happy (by a kind Providence or the muddling methods of Washington officialdom), for he was given a succession of agents each of whom seemed to be weaker and more inefficient than his predecessor. W. J. Pollock, who established the agency in 1878, was the only man of strong character to fill the office at Rosebud for a number of years, and his stay was brief, for he was soon promoted to the higher rank of field inspector in the Indian Service. When he left, he was succeeded by Cicero Newell, who took over the agency on May 3, 1879. Spotted Tail had one talk with his new official father, sized him up accurately, and pushed him to one side, the chief practically taking control of the agency and running it to suit himself.

Newell had been an officer of Michigan volunteers in the Civil War. Later he had been an office-seeker, holding such positions as town marshal of his home town, Ypsilanti. Old men at Rosebud Agency said that Newell had been a baker in Ypsilanti. On receiving his appointment as Spotted Tail's agent, Newell began recruiting a full agency staff among his Ypsilanti neighbors, and presently he started for the "Promised Land" of Dakota with a large caravan of Michigan families, including small children. All the men had appointments on the Rosebud staff, and the men and their wives were happy; but the way to Rosebud was long and wearisome, and by the time that they were set down on the west bank of the Missouri at the desolate Rosebud Landing, many of the party were homesick for Ypsilanti. From the landing they still had a hundred-mile journey by wagon across the trackless prairie before they at last came down into the hill-rimmed green bowl in which the agency buildings and the tipi camps of the Sioux were located.

Here they were in hostile country; for the old agency employees were waiting to welcome their new chief, and when they found that Newell had replaced them by giving their jobs to his Ypsilanti neighbors, there was almost a riot. Newell had antagonized nearly all the whites at the agency, and the Sioux—armed with Winchesters and Colt revolvers—looked far from friendly. It was at this moment, with the Ypsilantians huddled together and looking about them rather fearfully, that Spotted Tail came forward and took Newell and his flock under his protection. That was reassuring. By autumn the Ypsilanti families felt almost at home at Rosebud.

In September, Carl Schurz visited the agency, and Newell proudly introduced himself to the Schurz party as an Ypsilanti man. He then introduced the agency physician, from Ypsilanti; the chief clerk, from Ypsilanti; the carpenter, blacksmith, harness maker, warehouse clerk, all from Ypsilanti, Michigan. Carl Schurz, his eyes wide with wonder, led the agency physician to one side. "Tell me," he implored in a solemn whisper, "is there anyone left in Ypsilanti, Michigan?"

These Ypsilanti folk were about the strangest group that ever turned up at a Sioux agency, and as late as 1935 there were old men at Rosebud who still remembered and chuckled over the doings of Cicero Newell and his followers. The families had brought old feuds with them from Michigan, and as soon as they began to feel a little settled in their new surroundings, they split up into groups and, led by the ladies, revived the old bickerings of the Ypsilanti days. Agent Newell had several thousand Sioux to deal with—a full-time job for any agent—but he seemed to think that the petty quarrels among his Michigan followers were more important than any problem that had to do with running the agency. He spent a large part of his time going from one house to another, trying to patch up quarrels, mainly among the ladies. And while he thus labored, Spotted Tail sat in the agent's office, issuing orders. The Michigan people, particularly the women and children, were afraid of Indians, and at night any unusual noise in the Sioux camps that surrounded the agency threw them into hysterics. Agent Newell then went about from house to house, soothing the frightened families. He had considerable trouble in talking some of them out of a sudden desire to return to Michigan at once; and while he argued with them Spotted Tail sat in the agency office, running things to suit himself.

Agent Newell's main interest for several months was a plan for a bakery at Rosebud. It was his ambition to go down in history as the man who built the first steam bakery west of the Missouri in Dakota. He was soon busy, writing letters to the Indian Office on this all-important matter. The issue of flour rations to the Sioux was a waste of public money, he claimed. The Indian women had no stoves, no ovens; they mixed the flour into a mess and baked it in covered skillets in the embers of their campfires, misbranding the resulting product as bread. Newell told the officials in Washington that a bakery turning out real bread would improve the health of both whites and Indians at the agency and would save one-half the amount being spent to provide the Indians with flour. The bakery would thus pay for itself in a few months. This was the kind of plan the Washington officials liked. It was "progressive," and presently they authorized Agent Newell to use part of the fund provided for new buildings at the agency and to go ahead with the construction of his bakery. So Newell built a bakery, sent East for machinery, and trained some mixed-blood youths as bakery assistants. But when the great day came at last and the first batch of bread was produced in the new steam bakery, the loaves had the consistency and appearance of badly burnt bricks. Just what was wrong no one knew. A further effort produced another quantity of bread that was simply uneatable, and the Indians and whites at the agency then refused to give up any more flour and returned to their old custom of making bread in covered skillets. A year later a Dakota zephyr blew the bakery down, but no one seemed to mourn its passing.

Another of Newell's playthings was the sawmill. In the days when Spotted Tail's agency was in northwestern Nebraska the government had bought a sawmill, transported it at great expense to the agency, and started sawing lumber. Then the sawmill had vanished. It later turned up in the Black Hills, about one hundred miles away, where it was owned and operated at a good profit by a white man who was reticent as to just how a big sawmill happened to fly such a distance through the air and land in his lap. The government took no action, except to buy another sawmill for Spotted Tail. This mill was now at Rosebud, where it had been hauled in wagons from the Missouri, one hundred miles off. It had been unloaded and set up on a spot far from any timber to be sawed and far from any water for the steam boiler.

Newell wrote a report to Washington concerning this handsomely displayed and perfectly useless sawmill. He wanted to hire extra hands to move the mill to timberland on Little White River and put it to work. The Indian Office authorized spending the necessary funds. Then Newell had another idea—to change the mill from steam to water power by damming Little White River. Local men warned him that the stream could not be dammed, but he got the approval of the Indian Office and the funds to hire more men. He built a handsome dam; then a flash flood came, carrying away the dam and all the logs that had been cut in preparation for sawing lumber.

By this time even the trusting officials at the Indian Office were casting cold eyes in Newell's direction. He had squandered all the funds for building construction at Rosebud on his bakery and mill projects, he had increased the number of employees far above the legal limit, and there was nothing but failure to show for all this.[3]

[3] This account of Agent Newell's activities is made up from reports now in the National Archives, from items in the Yankton papers of the day, and from information obtained from white men and Indians at Rosebud. One informant was David A. (Jack) Whipple, who was employed at the agency off and on from 1877 and who had a clear memory of Pollock, Newell, Cook, and Spencer, the first four agents at Rosebud. One of these agents ran up big bills at stores in northern Nebraska, and when he was relieved as agent he took a train at a lonely way-station in the middle of the night and slipped away, leaving all the merchants in mourning.

In this year, 1879, Spotted Tail was giving an exhibition of what it meant for the Sioux to be free, controlled only by their own chiefs. At his agency the government farming program was being ignored, the Sioux attitude being that the government had taken vast tracts of land from the tribe and by treaty had agreed to feed, clothe, and care for the Sioux for at least one generation in payment for the lands taken. The Sioux said they had no need to farm, and that by trying to force them to work the government was violating the treaties. There was no effective Indian police force at Rosebud, Spotted Tail and his chiefs sticking to it that they had a right to police their own reservation. There was no jail and no real school, the Brûlés lumping jails and schools together in one category and rejecting them as things peculiar to the whites but unsuited to Sioux needs. What all this added up to was Sioux self-government at Rosebud, with Agent Newell immersed in his bakery and other projects and Spotted Tail sitting in the office and giving the Brûlés the kind of government they liked.

At Pine Ridge that champion of progress through strict discipline, Agent McGillycuddy, was loudly protesting over the shameful spectacle of a weak agent at Rosebud permitting his Sioux to do as they pleased. McGillycuddy had himself spread many of the choicest tales of the doings of Cicero Newell and Spotted Tail at that agency. He was enraged that such things could be at an adjoining reservation whence the poison of slackness might spread over into his Pine Ridge lands and corrode the splendid discipline which he was enforcing among the Oglalas. Over on the Missouri, northeast of Rosebud, Captain Dougherty was not worrying, like McGillycuddy, about what might happen if the slack discipline at Rosebud should infect his Lower Brûlés. It had already happened. Dougherty had organized an Indian police force in 1878, but his police were afraid of the Sioux and spent their time loafing about the agency buildings and drawing their pay. Early in 1879 he disbanded this force and recruited a new one, which he hoped would prove more effective. Then White Thunder, a leading chief of the Upper Brûlés of Rosebud, came to visit his Lower Brûlé cousins. Told of the agent's new police force, White Thunder expressed surprise that the Brûlés

Courtesy Smithsonian Institution, Neg. No. 3312-a

Fire-and-Thunder (left), Old-Man-Afraid-of-His-Horse (center), and Pipe (right) near Fort Laramie, Wyoming, 1867–68

Courtesy C. I. Leedy

Agent McGillycuddy (center) at Pine Ridge, circa *1884. With him are Sword, chief of police (seated, left); Standing Soldier, lieutenant of police; Billy Garnett, interpreter; and Young-Man-Afraid-of-His-Horse (seated, far right).*

of the Missouri had so little spirit that they would let a white man rule them by means of a group of renegade Sioux dressed up in soldier uniforms. The Brûlés at Rosebud would not tolerate such a condition. At Rosebud the chiefs were in control—a very proper arrangement. Were the Lower Brûlés women? At Rosebud there was no planting. Here at Lower Brûlé this white man was making the Sioux plant corn and hoe the patches. He—White Thunder—had come to invite all the Lower Brûlés to a big Sun Dance Spotted Tail was giving in midsummer, but, see—this white man was saying that none of the Lower Brûlés could leave their corn patches and attend the great ceremonies at Rosebud. If they tried to go, his police would stop them. White Thunder considered this very strange; he did not like it at all.

Neither did the Lower Brûlés, when their attention was thus concentrated on the matter, and one morning, soon after White Thunder had made his talk, a force of 150 warriors, mounted, painted for war and armed with magazine rifles, swept over the hills and down into the Lower Brûlé Sioux camps, where they raided the homes of all Agent Dougherty's new policemen. Whooping and firing their rifles, they ran all the policemen into the thickets; they then rushed back to the homes of these worthies (mostly log cabins), shot out all the windows, broke in the doors, shot the policemen's ponies and hogs, and rode off through the camps, firing promiscuously as they went. The next day the police came out of hiding and resigned in a body.

Captain Dougherty was very angry over this performance, but he had a sense of proportion and did not regard the naughtiness of his Brûlés as a sufficient cause for threatening them with war. He therefore left out all talk of calling in the cavalry, held a court of justice, and fined all the chiefs who were mixed up in the affair three months' sugar and coffee rations. The chiefs probably did not suffer, their admiring neighbors chipping in to furnish them with the necessary supplies to tide them over the three months. In June, nearly the whole Lower Brûlé tribe dropped their farming operations, just at the time when their little corn patches most needed attention, and all tracked off one hundred miles to Black Pipe Creek (near the present Norris), where they spent three glorious weeks at Spotted Tail's Sun Dance. They came home with many ponies, fine blankets, and other gifts from their Rosebud cousins.

Spotted Tail, when he planned his Sun Dance, was evidently unaware that he was going to be accused of spoiling the government plan to get the Sioux started at farming. Agent Newell, just arrived at Rosebud from Ypsilanti, Michigan, was also unconscious of this fact, and thus when Spotted Tail asked him to write notes to the other Sioux agents requesting them to permit their Sioux to attend the Sun Dance, Newell willingly did so. To Agent Schwan at Cheyenne River Agency, Newell wrote on May 19:

> You are requested by Spotted Tail to notify your people that there will be a sun dance at a place between Black Pipe and Black Creek next month, in the full of the moon. By doing so you will confer a favor on our Indian friends.[4]

Newell penned similar letters to the other Sioux agents. McGillycuddy of Pine Ridge probably passed on the news of the Sun Dance to his Oglalas. He was no believer in the Indian farming plan and had stated openly that farming was not possible on the Sioux reservation. He evidently did nothing to prevent his Indians' attending the Sun Dance. Agent Dougherty forbade his Indians to go; his tame Sioux at Crow Creek obeyed; his Brûlés at Lower Brûlé ignored his orders and went. At Cheyenne River, Agent Schwan read Agent Newell's letter and grew violently angry. He sent the Newell communication to the Indian Office in Washington with a peppery covering letter of his own, and the Indian Office promptly and severely reprimanded Agent Newell for assisting Spotted Tail with what it termed "a sun dance orgy."

Spotted Tail still did not seem to realize that anything was wrong. True, some of the agents had failed to co-operate properly in his effort to advertise the Sun Dance, but to rectify this failure, he sent out groups of his Brûlés to deliver in person his invitations at the several agencies. We have no reports from these Brûlé

[4] Letter books at Cheyenne River Agency. This Sun Dance was held in the western part of Rosebud reservation, near the present town of Norris.

parties, except from the group that went to Cheyenne River Agency. There Agent Schwan set his Indian police on them, and they were briskly run off the Cheyenne River reservation, a large number of shots being fired to speed them on their way. When Spotted Tail learned of this rude ejectment of his ambassadors, he dictated a letter to Secretary Schurz at the Interior Department (some white man in the Rosebud office wrote it for him), in which he denounced Agent Schwan, asked that Schwan should be removed from office, and made rude remarks concerning Captain Dougherty's conduct at Lower Brûlé. These two agents were military officers, and Spotted Tail wished them to be removed. "I have had enough of the military," the angry chief wrote. "I want my people to work. I want no more scouting. I have had mv belly full." Which meant that he wanted his Brûlés to work like white men (a fib) and that he wanted no Indian police, He meant police when he referred to scouting, having a clear memory of the Sioux scouts General Crook had enlisted in 1876–77.

Spotted Tail's performance had aroused the officials, both on the Sioux reservation and in Washington. The Brûlé chief was making the strenuous efforts of the government to control the Sioux look foolish by taking control from his weak agent and running things to suit himself. He refused to have a police force, and although he spoke nicely of wanting his people to work, none of them were working, and he had disrupted the very expensive farming program by taking thousands of Sioux away from their fields in June and keeping them for three weeks at Rosebud, happily helping at the Sun Dance.

The worst of it was that the Sioux were all siding with Spotted Tail as openly as they dared, making it very clear that they regarded the government Indian policy with loathing. That policy had been thought out by Eastern humanitarians, who had forced it on the government by pressure methods. The policy was most effectively in force at Cheyenne River, where Captain Schwan was vigorously controlling the Sioux and forcing them to work. Schwan had two types of Sioux: the wild Miniconjous and Sans Arcs, who had taken part in the war of 1876 and now had their camps far up Cheyenne River, where they kept away from the agent as much as they could and lived about as freely and contentedly as Spotted Tail's Brûlés were doing; and, along the Missouri and near the agency, the tame and friendly Sioux, mostly Two Kettles, who were getting the full force of Captain Schwan's attentions. He was forcing them to farm—or rather garden, for none of the families had enough land under cultivation to supply more than a small fraction of the food they consumed. Droughts and grasshoppers usually destroyed what little crops the Indians put in, but Schwan had a theory (approved by the humanitarians of the Indian Friends groups) that it was good training and discipline for the Sioux to work hard in their little fields, even if they had no crops to reward them. The Sioux did not appreciate that point of view and would have dropped their tools, only the agent would not let them; and he had his police to force them on, with the troops at Fort Bennett and Fort Sully to back him up when required.

In the view of the Eastern Indian welfare groups and the officials in Washington, conditions at Rosebud were shocking. Spotted Tail and his Brûlés were free, were doing as they pleased, and were happy. But there was no "progress," such as Captain Schwan was exhibiting at Cheyenne River. It was intolerable that Spotted Tail should entice Agent Schwan's Indians away from their fields and crops and encourage them to waste three precious weeks at his pagan Sun Dance rites; yet many of Schwan's Indians managed to evade his police patrols and go to the Sun Dance. Having enjoyed freedom for several weeks at Rosebud, part of them decided to stay permanently. Spotted Tail then ordered his weak agent to put the Cheyenne River families on the Rosebud ration rolls, but the agent was fearful about that and wrote to Washington for instructions. Carl Schurz (forgetting that he was now Secretary of the Interior, in charge of a policy of forcing the Sioux) gave way to his liberal inclinations, and on September 10 wired the Rosebud agent that it was proper to put Captain Schwan's runaways on the ration rolls.

Meantime there was another crop failure at Cheyenne River, which completely disillusioned the Sioux. At this critical moment groups of Brûlés from Rosebud came with legal passes that Agent

Schwan had to honor, and the visitors let out the fact that the Rosebud agent had orders to put Cheyenne River runaway families on the ration rolls. They also pictured the fine free life they led at Rosebud, and at once many Cheyenne River families began preparing to move. They were the Sioux who had felt the full force of Agent Schwan's policy of "progress" and of farming-without-crops. One group that was anxious to move was Chief Burnt Face's camp of Sans Arcs, who lived on Chantier Creek, only thirteen miles south of Schwan's office. They slipped away in a mass migration in October. They had few ponies, and most of them went afoot; but near the mouth of Bad River a group of Brûlés from Rosebud met them with ponies and wagons and guided them down into the "land of promise" at Rosebud. Agent Schwan was filling the air with angry shouts. He had his police on the run; he had the troops at Fort Bennett helping in the effort to round up his runaways; but on October 21 it was officially admitted that since the order to feed the Cheyenne River Indians at Rosebud had been issued, 264 of Schwan's Sioux had run away to Rosebud. This did not include the large number that had gone at the time of the Sun Dance and in July and August.[5]

Spotted Tail had undoubtedly won the campaign of 1879. He was a level-headed man, however, and no one realized better than he that he must not go too far in his opposition to the agents and to the government Indian policy. Yet it would be absurd to talk of his own policy and his statesmanship. He was an illiterate Sioux; his knowledge of the whites and their government was limited; and his policy, if we might dignify his actions by such a name, was one of pure opportunism. He understood that the Sioux were facing new conditions and that there was no way out of that, but he wished to have changes come slowly, and he had no patience with white officials who wished to rush the Sioux suddenly ahead into what they called civilization. In that Spotted Tail intended to oppose them, yet he told his people many times that they must never carry their opposition to the point of armed

[5] Cheyenne River and Fort Bennett letter books; Schwan to Commissioner of Indian Affairs, October 21; Spotted Tail to Carl Schurz, July 29. The Two Kettle fugitives from Cheyenne River remained permanently at Rosebud. They had a settlement north of the agency, near the present town of White River.

revolt. If they attempted to fight the whites, they would be destroyed.

When Carl Schurz and his party came to Rosebud in September, 1879, Spotted Tail and his chiefs were all friendliness. So was Schurz; yet he was worried over the entire want of progress at this agency. He had made promises to Congress, and he must show results. He therefore pressed the chiefs to induce them to cooperate in the matter of an efficient police force and to agree to start farming. Before Schurz came, Agent Newell had organized a police force. It was a nice-looking force, but there was something wrong about it. It was recruited largely from the ranks of the intelligent mixed bloods, regarded by the agent and other officials as superior to fullbloods. As police, however, the men just did not function. They spent their time loafing about their quarters or sitting on the bench outside, smoking and gossiping. Gradually the truth dawned on the officials, and presently they knew what any seven-year-old boy at Rosebud knew, namely, that mixed bloods were looked down upon with scorn by Brûlé fullbloods and that any mixed blood who tried to arrest a fullblood would be instantly lynched.

Shocked at Spotted Tail's duplicity, the officials now demanded a police force of fullbloods, but the chiefs stubbornly resisted such a change. It was in the autumn of 1879 that they gave in and a police force of fullbloods was recruited. The chiefs helped select the new police, and that was probably why the force failed to satisfy the officials. This was not a police force like that of Agent McGillycuddy or that of Agent Schwan, loyal only to the agent. The Rosebud chiefs had taken advantage of their unsuspecting agent to parcel out the new police among themselves, each chief selecting a man or two of his own, to represent his interests. Not a man on the force felt any loyalty except to his own chief.

Crow Dog was chief of police. He was almost unknown among the whites, but in the Brûlé tribe he had a reputation as a warrior and for intelligence. He was about forty-six at the time of his appointment as police chief. At his trial in 1882 a court reporter described him as a man of medium height, light complexion, long, glossy black hair falling to his shoulders and curling at the tips, clear and penetrating eyes, and a pleasing expression. He had a brother, Brave Bull, and a sister who were still living at Rosebud in 1885, and he belonged to a prominent family in the *Wablenicha* or Orphan band of Brûlés, formerly led by Chief Iron Shell and in 1880 by that chief's son, Hollow-Horn-Bear. Crow Dog lived in the camp of this band, north of the agency. He is said to have been the cousin of the gigantic warrior Yellow Horse, and also a cousin of Spotted Tail.

If the Brûlés had continued to lead their old free life of hunting and fighting, Crow Dog's cool courage and mental alertness would have won him high rank both in war and in the tribal council, but on the reservation time and opportunity had stopped dead, and Crow Dog had spent most of his days in hatching schemes with a group of kindred spirits who hoped to promote their own interests by undermining some of the chiefs, particularly the head chief, Spotted Tail. Why Spotted Tail accepted the appointment of such a man as chief of police is not clear, but it must be pointed out again that he as head chief was not in supreme authority, as many white persons believed. The decision to have a police force was made by the tribal council, and the chiefs in council then selected the chief of police and the policemen to represent the interests of all the leading chiefs and camps.

This undoubtedly undermined Spotted Tail's authority to a certain extent, but he was in no position to oppose the arrangement, and once it was made, he supported Crow Dog's authority as head of the police.

Crow Dog, once established as chief of police, seems to have begun to consider possible perquisites. Like Spotted Tail, he had been watching the operations of white men at the agencies for many years, and—again like Spotted Tail—he was able to think. Waiting for a favorable moment, he took a pair of his uniformed policemen and rode away into the vast emptiness of the eastern part of the reservation, where no Indians lived, but where white cattlemen from Nebraska were running large herds on the Indian lands. Neither the agent at Rosebud nor the officials in Washington seemed to be aware that white men were trespassing on the Indian lands to pasture their cattle on free grass, but Crow Dog knew of it and intended to look into the possibilities of the situation. So he and his two policemen rode eastward, and one night they came to a cow camp and had a meal with the cattlemen.

Crow Dog now opened the subject of his visit. The white men were grazing their cattle free on Sioux grass and it was only fair that they should pay the Sioux something for pasturage. He, Crow Dog, chief of police at Rosebud, had come to collect. The camp boss admitted that Crow Dog's reasoning was sound. But, as he pointed out to Crow Dog, he had already paid Chief Spotted Tail for the grazing privilege, and here was a paper, signed with the chief's name and bearing his mark, acknowledging receipt of $576 in full settlement for the season. Crow Dog scowled at the paper and said nothing. In the morning he and his two policemen had breakfast and rode away eastward. That evening they repeated their performance in another cattle camp, only to be met by a receipt signed and marked to prove that Spotted Tail had been ahead of them. And so it was at every camp. Crow Dog collected his two policemen and turned his pony's head toward home. He was angry, and when he reached the agency, he denounced Spotted Tail as a robber who was collecting money from white men for grazing and keeping the money. In effect, the chief was doing just what Crow Dog had intended to do. Crow Dog then announced that he was "looking for" Spotted Tail.

In this affair the conduct of both Spotted Tail and Crow Dog was no doubt culpable. In the chief's defense one might say that the Brûlé tribe had no funds and no need for them. It did need Spotted Tail, and if he served his people, he needed money. In the matter of the cattlemen he was simply imitating methods which he had learned from the whites—even from supposedly honorable United States Indian agents.

Visitors to Rosebud often saw long files of Sioux coming out of the hills and advancing in slow dignity one behind the other, wrapped in their fine dark blue and green blankets, their ornaments glistening in the sun. They were Sioux delegations from other agencies, coming to consult with Spotted Tail, and that chief had to entertain them hospitably and present them with gifts on their departure. This was one of the duties of the chief.

In the old free days he had had his own resources to enable him to keep open house for all visiting worthies and to present them with ponies, guns, or blankets on their departure. Now, at Rosebud in 1880, Spotted Tail was managing to keep up this state, and —strangely—the government that was trying to destroy the prestige of the chiefs was to some degree assisting him. Extra rations were still issued to chiefs, who had to entertain many guests, and the government was building a mansion for Spotted Tail on a little hill less than one mile north of the agency. Five thousand dollars were being expended on this work, the house being of two stories with eight rooms and a big council hall downstairs. It was larger and better built than the agent's house. Its construction had been ordered by Washington officials who were urging the agent to do everything possible to destroy the influence of chiefs.

The officials blew hot and then cold. They built a fine house for Spotted Tail, to increase his prestige among the Brûlés, and then wrote to Agent Newell, instructing him to report to the tribal council that the Great Father was very displeased with Spotted Tail's recent conduct. The result was growing confusion at Rosebud. Crow Dog and other men, thinking quite naturally that the officials had turned against the head chief, began to intrigue. They got up a petition against Spotted Tail and coaxed a number of chiefs and headmen into having their names signed to it. But before Agent Newell could send the petition to Washington, the chiefs held another council and changed their minds. When it came to the touch, they frankly realized that there was no other chief in the tribe who had the necessary stature to take Spotted Tail's place.

The officials at the Indian Office were still puzzling over just how Agent Newell had managed to spend all the Rosebud funds and why he had so many extra workers on his rolls, when in the late summer, 1879, an angry lady at Rosebud wrote them a very enlightening letter. She was a Mrs. Owen of Ypsilanti, Michigan. Her husband, Dr. Owen, had been given the appointment as agency physician at Rosebud by Newell and had come to Dakota with the agent. In July, Newell had written to Mrs. Owen at Ypsilanti and offered her a clerkship at a handsome salary. She closed her home in Ypsilanti and came to Rosebud at considerable expense, only to find that Agent Newell seemed to be strangely reluctant to execute his promise and give her employment. Visiting her several Ypsilanti friends at the agency, she was told by them in confidence that Agent Newell never gave appointments, that he invariably expected to be paid in advance, fifty dollars for a minor position, one hundred dollars for a better-paid one. It was then that Mrs. Owen sat down and wrote a very angry letter to the Indian Office.[6]

So (thought the officials) this was why Agent Newell was always planning new projects and hiring large numbers of extra workers at his agency. Inspector Pollock was sent to Rosebud and reported that most of the charges were true. The Indian Office hushed the scandal up and did not even inform Newell that he was to be removed. He learned the fact from an item in a Yankton newspaper just after New Year's, 1880, and took what countermeasures he could. Thus in early February the Rosebud chiefs sent a petition to Washington, stating that they wished to keep Newell as agent, and on February 7, Spotted Tail wrote to President Hayes to the same effect. That month Rosebud was humming with excitement, the tribal council meeting frequently and the chiefs having their names signed to petitions, some of which were intended to keep Newell as agent, others containing complaints against Spotted Tail. To add to the excitement, there was a gold rush on. Rosebud lies in the plains and has clay and

[6] Newell denied these charges and demanded a hearing from Secretary Schurz; but Schurz ignored him, and in the printed reports for the year the whole affair was hushed up and no explanation given for Newell's removal. In the National Archives there is a letter from Dr. N. Webb of Ypsilanti, Michigan, to Senator H. P. Baldwin, February 2, 1880. Dr. Webb was the father of Mrs. Owen, and he states that when his daughter wrote to Washington, Newell brought a charge of theft against Dr. Owen. Owen, his wife, and their two children then left the agency to return to Michigan. Halfway to the Missouri River they were overtaken by Indian police Newell had sent after them, and the doctor was arrested and his property seized. The doctor's wife and two children were left stranded in the prairie. At the agency Newell accused Owen of stealing the government surgical instruments, took the instruments from him, and then let him go. He had a dreadful time getting to Yankton, where he found his wife and children living on charity. The family did not have a penny, as Newell had kept the doctor's salary from him. Dr. Webb in his letter stated that the surgical instruments were his own. He had given them to his son-in-law when he was appointed physician at Rosebud and everyone in Ypsilanti knew that fact.

sandy soil. As far as is known, no one ever found gold nearer than the Black Hills, about two hundred miles westward. But in 1879 there had been two or three survey parties working on the reservation, escorted and guarded by Indian police; the Sioux did not understand the purpose of the surveys, and a rumor had started that the whites were either seeking gold or had actually found it. The result was an astonishing outbreak of gold fever among the Sioux, who began to quarrel violently over claims to tracts of land they thought might contain some of the precious metal. Crow Dog became involved in these gold-claim feuds; a majority of the Indians turned against him, and he was suspended from his position as chief of police, evidently at the request of the tribal council. He blamed Spotted Tail for this misfortune and redoubled his intrigues against the head chief.

Newell was succeeded as agent by John Cook, another political appointee, on April 3. Like Newell, Cook had been an officer during the Civil War; nevertheless, he was a man of weak character, one contemporary report describing him as a long-haired effeminate person. Spotted Tail, after one talk with him, said publicly that Cook was a small man, of no account. This was the signal for Crow Dog and others who were opposing Spotted Tail to get up a petition stating how very much they liked Agent Cook, and they induced a number of chiefs to let their names be set down on the petition. Naturally, Agent Cook began to show favor to Crow Dog and made it fairly clear that he did not like Spotted Tail.

We now come to that stirring event, Spotted Tail's raid on the Carlisle Indian School in Pennsylvania. About two years before, the leading thinkers among the Eastern Indian Friend groups had decided that education could solve the Indian problem in just a few years if Indian children were removed from the evil influences of family and tribal relationships and secluded in big government boarding schools, far from the reservations. Secretary Schurz was all for trying this experiment. He obtained from the War Department the abandoned cavalry barracks at Carlisle, and Captain R. H. Pratt of the army was appointed head of the new Carlisle School. Then, in late summer of 1879, Pratt came out to the Sioux reservation to recruit his first pupils. He went to Rosebud to start, where he received a severe shock; for it soon became apparent that the Brûlés of Rosebud were not at all interested in education, certainly not that form which required them to give up their children and see them carried off to an unknown place far away in the white people's country, perhaps never to return. They did not regard such a plan as noble. It was in their view about as wicked as placing their children in an open boat and letting them drift off down the Missouri River, to meet an unknown fate.

Appealed to for aid by Pratt and the local Protestant Episcopal missionary, Spotted Tail now offered to send four of his sons and some of his grandchildren to the new school, and the moment that he made the offer, the other chiefs and leading Indians crowded up to offer their sons. Pratt got thirty-four pupils at Rosebud. Going on to Pine Ridge, he obtained further recruits, despite the fact that Red Cloud stood like a rock in opposition to the school plan. In all, Pratt obtained at Rosebud and Pine Ridge eighty-four pupils (sixty boys and twenty-four girls) to form the nucleus about which he hoped to build up a great Indian school.

At this moment the leaders of the Eastern Indian Friend groups gushingly termed Spotted Tail's giving up his sons and grandchildren to Pratt the noble act of a great and wise chief. But Spotted Tail had no liking for schools, and he probably had a very prosaic, not to term it mercenary, reason for coming to Pratt's assistance. Perhaps his action was due to the fact that Pratt offered a good position and salary to the chief's white son-in-law, Charles Tackett.

Turning back to 1875, we find that Spotted Tail in that year was coaxed by some Roman Catholic nuns who visited his agency into agreeing that they should take his favorite daughter and educate her in their convent. The girl, however, had her own view of this plan, and on the very day that she was to be turned over to the nuns she eloped with a young warrior and went to live in a wild camp in the Powder River country, where her father could not touch her. Later she came home, disillusioned, and divorced her Indian husband, which probably meant that she chased him out of the tipi, throwing all his personal belongings after him. She had then married Charles Tackett, who was living at Rosebud. When Pratt came to Rosebud and the Indians refused to let him take any of their children, someone suggested that the Indian parents would be more willing if Pratt engaged Tackett and his wife to go to Carlisle with the children, watch over them, and report to Rosebud from time to time. Pratt accepted this idea, and it was then that Spotted Tail came forward and offered his sons and grandchildren.

In 1879 Carlisle School was the pet project of the Indian Office and was under the special patronage of Carl Schurz, secretary of the interior. It was the particular love of the Eastern Christian and benevolent groups who were interesting themselves in Indian welfare, and all these people had high hopes that Captain Pratt was beginning a movement that would very soon create a new and glorified Indian race, educated and brought up level with the whites in all respects. The Indian Office officials were so proud of the school that by New Year's, 1880, they were planning a gathering of notables at Carlisle in the following June, to celebrate the completion of the first year. They wrote to the agents at Rosebud and Pine Ridge, sending carefully chosen lists of chiefs who were to be brought East at public expense to attend the Carlisle meeting.

While the officials made their plans, Spotted Tail began to have his doubts about this school. During the winter Tackett wrote to the chief that all thirty-four of the Rosebud boys and girls at Carlisle had been baptized and taken into the Protestant Episcopal communion. The children had been given Christian names. Spotted Tail's oldest son at the school had been renamed William; one of the younger boys had been named Pollock, after the first Rosebud agent; and a third had been named Max. If this news was supposed to please the great chief, it only showed that Captain Pratt at Carlisle, like the officials at the Indian Office, knew very little about Spotted Tail. He was angry. He was a Sioux pagan and proud of it, and he wished his boys to follow in his path. He had trusted his boys to Pratt, and that man had put pressure on the boys and forced them to change their faith when their father was not present to advise them.

The Indian Office went on with its plans for the Carlisle gathering, to be held in June, 1880. It sent a list of chiefs to be brought East, as early as January, the chiefs being the heads of the bands at Rosebud. On this list were Spotted Tail, Two Strike, Swift Bear, Red Leaf, White Thunder, and "Old Man Crazy Horse," the father of the famous Oglala war chief of the same name. The agent protested that Black Crow should be included, even if he were only a headman, for he had given his only daughter to Pratt to be taken to Carlisle. In the end Swift Bear and old Crazy Horse withdrew, because of their age, and Black Crow and Iron Wing went in their places. These Rosebud chiefs and the ones from Pine Ridge reached Carlisle ahead of time and spent some days living at the school and inspecting it.

Captain Pratt and the Washington officials were so blinded by their own enthusiasm that they could see no flaw in the Carlisle establishment, and they could not even picture anyone coming to inspect the school with critical eyes. Yet that was exactly what the Sioux chiefs now did, and almost at once they were at daggers drawn with Captain Pratt, who would brook no criticism. Pratt had a notable temper, and when he loosed it, he expected any opponent to wilt at once. As Spotted Tail was the principal critic of the school, Pratt forgot his manners and began to storm at him, and the next thing he knew he was almost blown away by the counterblast of the Brûlé chief. By the time the other distinguished visitors (including Bishop Hare of the Protestant Episcopal church and a group of prominent Quaker ladies) arrived, Captain Pratt had had more than enough of Spotted Tail. The Brûlé chief now put on his company manners and was very polite to the Bishop and the Quaker ladies. He was photographed with the ladies, who seemed to be quite pleased, and one wonders if anyone had told them that at the moment their dear friend from Rosebud had four wives and was courting a Sioux girl he intended to make number five. All went well until Spotted Tail's white admirers insisted on his making a speech to the assembled notables. He stood up and began to speak—one sentence, then a halt while his interpreter turned his words into English, then another sentence—and he had not spoken five sentences before the assembled guests began to look shocked. They had expected the chief to praise Captain Pratt and his school; but he was criticizing both most severely. He said angrily that Pratt had made "a soldier place" of Carlisle. He had all the boys in military uniform, drilling with guns and under strict military discipline. He was turning the sons of chiefs into common workingmen, training them in farming, carpentry, and bricklaying. This was all wrong. The boys and girls had been sent to this school to be taught to speak, read, and write English, and to learn other matters out of books. Before this speech was concluded, it was glaringly apparent to everyone present that Spotted Tail's and Captain Pratt's ideas of the aims of education were widely at variance and that Spotted Tail did not intend merely to talk about this. He was going to act.

Pratt was undoubtedly very much relieved when the chiefs left Carlisle to go on to Washington for conferences with the officials. The Captain sat down and wrote worried letters to his friend Major Andrus in Washington, urging that when the chiefs returned home, they should be taken by a route that would avoid his school. But Spotted Tail had his own plans, and he brushed aside those of the Washington officials and came back to Carlisle, bringing Red Cloud and his chiefs with him. The chiefs wished to have it out with Pratt in the presence of all the Indian pupils. Pratt objected, but the chiefs blew his objections away, and the Captain ordered the children assembled in the school chapel, perhaps in the hope that the sacred character of the place would deter the chiefs from taking any violent action. A vain hope; for now all the leading chiefs, including Red Cloud and Red Dog, made angry speeches, offensive to Captain Pratt and derogatory to Carlisle School. The blow then fell. Spotted Tail announced that he was taking all thirty-four of the Rosebud boys and girls home with him. That meant the end of Carlisle School. Its back would be broken.

Pratt sent frantic telegrams to Washington. Secretary of the Interior Schurz wired Spotted Tail, forbidding him to take the Rosebud children, and this word from the great official caused some of the chiefs to desert Spotted Tail. But the angry Brûlé said that if the others were too weak to act, he was not, and he was taking his own children, in the face of the opposition of every official in Washington. He then gathered up his sons and grandchildren and hung grimly to them, while Captain Pratt moved heaven and earth to make him let go. Red Dog, following Spotted Tail's lead, took possession of an Indian girl (probably his granddaughter) and insisted on taking her back to Pine Ridge. Pratt could not handle this situation, and Major Andrus was rushed from Washington to aid him, but Spotted Tail defied Andrus as he had already defied Pratt. In one last effort to stop the outrage, Secretary Schurz telegraphed from Washington that if Spotted Tail took his own children from the school, he would have to pay their transportation and other charges home, for the government would not pay one penny. The Brûlé chief ignored this and held on to his children.

Thus the Sioux chiefs left Carlisle—left it shuddering and shaking from the impact of their actions. Major Andrus accompanied the Indians to the railway station, still trying to persuade them to give up their dreadful purpose. Pratt stayed in the school to herd the rebellious Indian children; but some of them slipped away and smuggled themselves onto the train. They were found by Andrus and dragged off weeping. Old Red Dog was weakening, and just before the train left, he permitted Andrus to remove the small girl relative he had taken from the school. The train started; but at Harrisburg it was searched again and another Indian girl runaway was found and carried back to Carlisle.[7]

Spotted Tail was never forgiven by the officials and the leaders of the Indian welfare groups for this deed. He had struck a savage blow at that new panacea, education, which they believed was going to transform the Indian race. Pratt spread it about that the chief had acted for mercenary reasons, that he had demanded forty dollars a month increase in the pay of his son-in-law Tackett and had taken his children from the school when this favor was

[7] This account of Spotted Tail at Carlisle comes partly from information obtained from old men at Rosebud; but in the main from notes which my friend Harry Anderson obtained for me in the National Archives, where the correspondence connected with this affair is preserved.

refused. The Brûlé chief may have said something about an increase in Tackett's salary, but it was apparent that his actions at the school were caused by his disapproval of Pratt and his purposes. Spotted Tail could see the fundamental wickedness of Pratt's plan to take Indian children hundreds of miles from their homes, confine them, and root out of them every trace of their Indian origin. What was really strange was that kindly Christian people in the East heartily approved of Pratt's plan. They knew that he had men and women on his payroll under the euphemistic designation of disciplinarians whose main duty was to thump recalcitrant Indian boys and girls into submission. When some years later a bill was introduced into Congress to put a stop to beatings in Indian schools, they sided with Pratt. He went to Washington and stormed about, declaring that this bill, if it became law, would mean the end of Indian schools. He won his fight, and it required another twenty-five years for the government to advance to the point where it could run an Indian school without including thugs on the staff whose duty it was to beat the pupils.

The officials hushed up as far as was possible Spotted Tail's raid on Carlisle, but they were upset and angry, and even after the chief got home, the Indian Office sent him a telegram urging him to alter his decision and send his children back to Carlisle. He would not, and the great majority of the Sioux backed him up. When Captain Pratt sent men to the reservation to recruit more boys and girls for his school, they could not find any Sioux parents who were willing to give up their children. The Indian agents did all in their power to help—coaxing, bribing, and making threats—but all in vain. Even the blindest of the Eastern Indian Friends now had to abandon their attitude that the Indian parents longed to send their children to school, and presently a system of compulsion was devised, and every Indian agent became a professional kidnapper who sent his Indian police out to ambush and drag in weeping Indian boys and girls, to be sent off under guard to one of the big government schools.

The officials were trying to make Spotted Tail pay for what they regarded as his outrageous conduct at Carlisle. When he got home, the Indian Office sent a telegram to Agent Cook, instructing him to inform Spotted Tail and the tribal council that the Secretary of the Interior was very displeased with the chief. Whether so meant or not, this was an invitation to Spotted Tail's rivals in the tribe to resume plotting against him, which they did at once. But they had bad luck. They could plan and hold councils and turn the Brûlés for the moment against the head chief, but there was not a strong leader among them, and when anything went wrong the entire tribe turned to the great chief for aid. This was what happened in late July, 1880.

While Spotted Tail was in the East, discipline had slackened at Rosebud. For ten years he had taken precautions to prevent any raiding, but while he was at Carlisle six young Brûlés slipped away and made a horse-lifting raid into northern Nebraska. They got some horses and killed a ranchman. All the Nebraska border was in a flame of resentment and demands were being made that the raiders should be caught and tried. Agent Cook suggested to Crow Dog, chief of the Indian police, that he should attempt to arrest the culprits, but Crow Dog hesitated. He said that these young men were very popular in the tribe, with strong backing, and if the police tried to arrest them, there would be a big fight. Crow Dog was, of course, the leader in the plan to pull Spotted Tail down, and here were his credentials. He had neither the force to fight nor the ability to persuade. Spotted Tail came home and was appealed to, for this crisis was serious and the raiders had to be arrested to satisfy the outraged citizens of Nebraska. The chief went to the camps alone, talked to the young men and persuaded them to come with him to the agency, where they submitted to arrest and were sent under guard to Fort Randall on the Missouri, to be held for the Nebraska authorities. To complete a fine piece of work, the head chief now took up a collection and sent a check for $332.80 to Carl Schurz, to be used in employing counsel to defend the Brûlé youths. When informed from Washington that the sum was not enough, Spotted Tail took up a second collection. There was not another chief on the Sioux reservation who could have shown this decision and skill in action. This was the first instance in history of a Sioux chief's transmitting money by check. He signed the check with his mark.

Both the Brûlés and the officials were pleased with Spotted Tail's action, but later in the year he was in disgrace again. He picketed the traders' stores at Rosebud, employing the Indian police for this most irregular purpose. What his object was is not clear. He was evidently bent on collecting money from the traders, perhaps to reimburse himself for the expense of bringing his children home from Carlisle, when Carl Schurz had refused to pay the cost from government funds; or was this an effort on the chief's part to increase the fund for the defense of the six Brûlé raiders? Whatever his object was, he succeeded in enraging the traders and Agent Cook, and he turned the tribal council against him once more.

Cook heard an uproar and went out of his office, to find a mob of Sioux gathered in front of the traders' stores. Going across with his interpreter to investigate, he found the Indian police at work preventing any Sioux from entering the stores to trade. He ordered the police to go away, and they told him through the interpreter that they were here on Spotted Tail's orders and would obey no one else. A fierce warrior, Thunder Hawk, who was feared by most of the other Indians, now took the agent's side and forced the police to go away. Informed of this, Spotted Tail mounted and rushed from his camp to the agency, where he stormed in on Cook and ordered him to keep his hands off the police, who were under his own orders as head chief. Pushed on by the angry traders, Cook for once had the courage to defy the chief. He called a tribal council, and the council was persuaded to decide that the police were under the agent's orders and that Spotted Tail had no right to employ the police for his own purposes. The police force was now disbanded and a new force recruited. Crow Dog had lost his place as chief of police, and he bitterly blamed Spotted Tail.

That winter, 1880–81, Crow Dog was busy with his plans for destroying the head chief. He had two letters written, charging Spotted Tail with misconduct, and he coaxed some of the chiefs into having their names written down as co-authors of the charges. They were minor charges; but the fact that Agent Cook would accept such letters and transmit them to Washington indicated that Spotted Tail no longer had the good will of the officials.[8] He was left open to attack from malcontents in the tribe, and Crow Dog and his friends took full advantage of this.

Spotted Tail was now living in his own camp north of the agency. He had four wives and a separate tipi for each wife and her children. The chief had tried living in the mansion the government had built for him but had found it too uncomfortable. No furniture had been included in the government's handsome gift, and the Spotted Tail family found camping out in the big bare structure far less to their taste than living in tipis. Besides, try as they might, Spotted Tail's wives could not learn how to cook and work in a house. They did not even solve the problem of cleaning, and the new dwelling soon grew too dirty to bear.

Spotted Tail was in love again, and early in 1881 he "stole" the wife of a Brûlé warrior. She was young and handsome, and the chief was quite pleased with his new acquisition, even if his four wives were upset. Pretty girls had always been his one indulgence. He did not smoke or drink; but girls he could not resist. In an affidavit concerning cattle, made in 1879, Spotted Tail stated that he was known among the whites by that name but that his own people called him Speak-with-the-Woman, and to speak to a woman meant in Sioux to make love to her.[9] The Washington officials and leaders in Indian Friends groups, who admired Spotted Tail and called him a hero and great leader, knew about his numerous wives and his penchant for girls, but averted their faces and said nothing. The Brûlés at Rosebud knew much more about these matters and did not avert their faces. Chiefs with several wives and girl-stealings were nothing out of the way; the Brûlés gossiped and laughed over Spotted Tail's latest conquests. Many Sioux of the present day, however, take a highly Christian view and condemn Spotted Tail. They have forgotten that plural wives used to be the tribal custom, that girl-stealing was a tribal

[8] The letters accusing Spotted Tail are in the National Archives. As far as I can find no action was taken in Washington, and the letters were not even given to the newspapers.

[9] Affidavit in the case of the United States *v.* Isaac Coe, Levi Carter, and John Bratt, quoted in H. B. Paine, *Pioneers, Indians, and Buffaloes* (Curtis, Nebraska, 1935), 96.

Courtesy The New York Public Library

Spotted Tail, Captain Pratt, and Quaker ladies at Carlisle School, June, 1880.

Courtesy National Archives

Brûlé Sioux boys at Carlisle School, June, 1880. Standing, left to right: David Blue Tooth, Nathan Standing Cloud, and Pollock Spotted Tail; seated, left to right: Marshall Bad Milk and Hugh Whirlwind Soldier (grandson of Spotted Tail).

sport, and that the very first white man who recorded a meeting with the Brûlés (the trader Jean Baptiste Truteau, coming up the Missouri in a boat in 1794) set down the fact that the Brûlé camp he met was in a continuous uproar over the stealing of horses and wives. From the Christian point of view it is impossible to defend Spotted Tail's attitude toward women; but from the old-time Sioux angle we may say that this chief was never known to injure a woman, that he doted on his wives, being as fond of the oldest of them as of the newest recruit, and that he was a fond parent to his numerous children, who all worshipped him. It may also be noted that army officers and their wives, who were not in the habit of associating with reprobates, regarded Spotted Tail as a man of the highest character, welcomed him at their homes, and permitted their young daughters to go to the camp to visit Spotted Tail's wives and children.

In 1881, nothing appears to have been said about the girl Spotted Tail had taken. The modern Sioux have built her up into the main reason for Crow Dog's act in August, 1881. Their story is that the girl was beautiful, almost as handsome as Black Buffalo Woman, who was stolen from her husband by Crazy Horse in 1870. The girl of 1881 lived in the camp of Chief Lip on Pass Creek, far from Rosebud, but her father sold her to a man named Thigh (sometimes called Medicine Bear) who brought her to a camp near the agency, and there Spotted Tail saw her. She was easily persuaded by the elderly but still handsome head chief, and soon moved to his camp. At the time there was no talk, but the Sioux of today claim that Crow Dog took up the cause of the injured husband, who they explain was crippled. Crow Dog and his friends paraded publicly to the head chief's camp and made fine speeches, appealing to his better self. They offered ponies and other gifts if he would return the girl, but he refused.

On the Fourth of July all the Brûlés were at the agency for a big celebration, and Spotted Tail and Crow Dog came together in the open space between the big government commissary building and Trader Clark's store. Some angry words were exchanged, and then Crow Dog pushed the muzzle of his Winchester against the chief's breast, shouting fiercely that now he would kill him.

Spotted Tail stood looking at him steadily. "Why don't you shoot?" he asked. "If I had a gun on you I would certainly shoot." But there were too many of the head chief's friends in the crowd of Indians, and Crow Dog backed away.

These modern Sioux stories, particularly the one concerning this girl, do not ring true. The reports made immediately after the murder make it clear that this was a political assassination, planned carefully by a little group of ambitious Brûlés whose object was to place one of their own number in Spotted Tail's position as head chief. The plotters were much encouraged when Spotted Tail lost the support of Washington officials over his raid on Carlisle School in 1880. The message sent to the Rosebud Indian council by the Secretary of the Interior, expressing his high disapprobation of the head chief's removal of his children from the school, appears to have heartened the plotters. They found, however, that the old chief still had strong support in the tribe. They dared not act. Then the matter of the Ponca lands came up and the officials in Washington asked the Rosebud agent to send a delegation of chiefs East, to make an agreement returning the lands. Crow Dog and the other plotters now decided to prevent Spotted Tail from going with this delegation and re-establishing himself in the good graces of the high officials.

In 1877 the Ponca lands had been given to Spotted Tail's Brûlés, who refused to live on the lands and went to Rosebud; but the Ponca reservation was still Brûlé property by legal agreement. When citizens of Nebraska took the part of the unfortunate Poncas, a judgment against the government was obtained. It was now necessary to return the Ponca lands, but the officials could not do that, as they had legally handed over the lands to the Brûlé Sioux.

Carl Schurz was in a most painful position. The Ponca removal to Indian Territory had been completed before he became secretary of the interior, but—urged on by the fiery eloquence of young Helen Hunt Jackson—the Indian Friends groups were assailing Schurz with a vindictiveness that was only equaled by their utter lack of fairness. Schurz, honest and kindly, was being crucified by his own friends among the so-called liberals, who were so blinded by passion that they refused to see that as secretary of the interior Schurz could not brush aside all law and act instantly in restoring the Ponca lands. The only practical course was to induce the Brûlé Sioux to give back the lands by legal agreement. To the little group of Brûlé plotters this Ponca crisis was an unexpected setback. No one knew better than they did that when trouble came both the officials in Washington and the Brûlé tribe turned to Spotted Tail for help. Now they had to act or give up their plans for making an end of the head chief they hated.

On August 5, 1881, the Brûlé tribal council met at Rosebud Agency, and under the leadership of Spotted Tail the chiefs and headmen decided on the terms to demand of the officials for the return of the Ponca lands. They also voted to select the men who were to go to Washington, and (as the plotters had feared) chose Spotted Tail as head of the delegation. After the council had ended, a feast was held, and then the large body of Sioux that had come to the agency scattered and started to return to their camps.

Crow Dog—having lost his position as chief of police—was picking up some money by selling firewood (cut by his wife) at the agency. On the day of the council he had brought a load of wood to the agency and delivered it. He and his wife then started back to their camp north of Rosebud, taking the trail that led past Spotted Tail's deserted mansion. They had removed the box from the wagon and were riding on the running-gear. Beyond the Spotted Tail house they saw a lone rider coming up behind them from the agency. Crow Dog handed the reins to his wife and jumped down, kneeling on the trail. Spotted Tail, coming along at a brisk speed, saw someone kneeling in the dust, probably tying his moccasin string. He came on. When he was almost up with the wagon, Crow Dog pulled a Winchester out from under his blanket and fired. The chief fell off his pony, throwing up a great cloud of dust as he struck the trail. He staggered to his feet, drawing his revolver, but before he could fire he fell and lay still. Crow Dog leapt back on the wagon, whipped up his ponies, and fled to the safety of his own camp.

Henry Lelar, chief clerk, who was in charge of the agency, sent an Indian eyewitness account of the killing to the Indian Office on August 5. This Sioux account differs from others in stating that Crow Dog was driving toward the agency, not away from it, when he encountered Spotted Tail. He says:

We had a council and a feast, after which Spotted Tail mounted his horse and started home. The council had broken up and the people were scattering out. Spotted Tail was in advance. I saw Crow Dog coming toward us in his wagon. He had his wife with him. He got out of his wagon and was stooping down. When Spotted Tail rode up to him he suddenly raised up and shot him through the left breast. The chief fell from his horse but at once rose up, making three or four steps toward Crow Dog, endeavoring to draw his pistol. He then reeled and fell backward, dead. Crow Dog jumped in his wagon and drove off at full speed toward his camp, some miles away.

Three chiefs (Two Strike, Ring Thunder, and He Dog) were coming along the trail behind Spotted Tail, all on foot. They and other witnesses spread the alarm, and bands of mounted warriors came pouring out of all the camps, most of them heading for Spotted Tail's camp or the agency. Agent Cook was away; the Indian police made no move; and Clerk Henry Lelar, who was in charge, was too stunned by the sudden tragedy to know what to do. It looked as if fighting was about to break out between the enraged factions, but the Sioux had a horror of killing their brother Sioux, and peacemakers were soon busy, trying to force a settlement. The Crow Dog faction was offering blood-money, and the bewildered family of the dead chief was induced to accept. They took six hundred dollars, a number of ponies, and some other articles in payment for the wanton murder of one of the greatest Indian chiefs that had ever lived.

Henry Lelar was not satisfied with this Sioux settlement of the murder. He now had information that the killing was the result of a deliberate plan hatched by a few conspirators of desperate character, among whom Crow Dog was the leader. Crow Dog, Lelar stated, had the intention of winning for himself the rank of head chief; other informants stated that Black Crow had been selected by the plotters to succeed Spotted Tail and that Crow Dog was merely the trigger man. The plotters now had surrounded themselves by a large body of armed followers and were loudly declaring that they would fight any group that interfered with them. Lelar called a council of the chiefs on the day after the murder; the chiefs then warned the plotters in the name of the tribe that they would not be permitted to make further trouble—trouble which would probably end in a large force of cavalry taking control at Rosebud. Crow Dog and Black Crow were then talked into submitting to arrest and were taken by the Indian police to Fort Niobrara in northern Nebraska and placed in the guardhouse.

Brought to trial at Deadwood in the Black Hills in 1882, Crow Dog stood alone in the dock as the murderer. He was convicted and sentenced to hang; the case was appealed, however, to the United States Supreme Court, which handed down a decision that the Dakota courts had no jurisdiction over crimes committed by Indians on Indian reservations and that Crow Dog had settled the murder by tribal custom when he and his friends had paid blood-money for the slaying of Spotted Tail. He was therefore set free, to return to Rosebud and resume his plotting.

The murder of Spotted Tail proved a great disappointment to the men who had planned it. They were what the Sioux called small men: not one of them of proper stature to stand in the place of the great chief they had done away with. There was not a man in the Brûlé tribe fit to take his place. His own son, Young Spotted Tail, was only fit to pull the trigger of a Winchester; the other aspirants, Black Crow and White Thunder, were just average minor chiefs. After the killing, these small Brûlés started lively intrigues, striving to win support that would gain them the coveted rank of head chief. Black Crow had no real chance. Young Spotted Tail followed Crow Dog's method and shot his rival, White Thunder, to death.[10] That cost him most of his friends.

[10] This eldest son of Spotted Tail does not seem to have been given a Christian name. The Sioux called him *Sinte Galeska Chika:* Little Spotted Tail. He signed the land agreement of 1882 simply as Spotted Tail. The Brûlés gave him the nickname of "Tin Cup." In 1882 he was attempting to gain his father's position as head chief, was living in his father's mansion, wearing citizen clothing, and driving about in a nice buggy. The Brûlés would not accept him as head of the tribe, and the government from the first had no intention of recognizing another head chief at Rosebud. Spotted Tail finally went to live on a land allotment

Slowly the awful truth found its way into the brains of the Rosebud Sioux. The government would never recognize another head chief, and by killing Spotted Tail the plotters had removed the one man the Brûlés had who could stand between them and the government's efforts to break their tribe to pieces and place them completely under the control of white agents.

Spotted Tail was given a pagan Sioux funeral. Laid out in his fine buckskin clothing, ornamented with brightly colored porcupine quillwork, his face painted red, a piece of suet between his lips, he was mourned by his family. Later his wives permitted the missionary to place his remains in the Episcopal cemetery, where he still lies under a white marble monument, and near him is buried his old friend James Bordeaux the trader.[11]

---

south of the agency, where the New Deal in the 1930's built Spotted Tail Park on young Spotted Tail's land, evidently with the mistaken idea that this was the great chief's land. It is now forgotten. Some writers have imagined that this eldest son of the old chief was the man later known as William Spotted Tail; but William was the boy, aged eighteen, who was sent to Carlisle in 1879 and was put in the harness shop to learn that trade. Young Spotted Tail was at Rosebud all the time when William was at Carlisle. William became the head of the Spotted Tail family and was succeeded by his son, Stephen.

[11] This narrative of the killing of Spotted Tail has been put together from official documents, now in the National Archives, and some accounts obtained from old men at Rosebud. Knife Scabbard, Sore Eyes, and David Murray of Rosebud gave information. The story of the girl who was taken by the chief in 1881 came from Edward Herman of Rapid City and was confirmed by Joseph Eagle Hawk and John Colhoff of Pine Ridge. The modern Sioux are inclined to pretend that the stealing of the girl was the one and only cause of the murder. They forget that Crow Dog was openly threatening to shoot the chief long before the girl appeared on the scene and that, as soon as he had killed Spotted Tail, Crow Dog lost interest in the girl and her injured husband. The Sioux story is that on the day Spotted Tail was killed an honored and venerable chief (we will not give his name, but we have it) drove to Spotted Tail's camp in his buggy and told the girl that Spotted Tail's wives and family would kill her. He offered to take her to a place of safety, and she got into the buggy. The old rascal took her to a distant camp and forced her to live with him; but he had several wives at home, and after a fortnight or so, he returned to his own camp. The girl went back to her father in the camp of Chief Lip, and her father promptly sold her to an old and cruel husband, from whom she presently ran away. Handsome girls with no strong protectors had a hard time among the Sioux in 1881.

3
Red Cloud as Lucifer

AFTER TWO crop years Carl Schurz had almost no results to exhibit as the fruits of his farming crusade on the Sioux reservation. He had pledged his word to Congress that he would advance these Indians swiftly toward self-support, but they were still living on a basis of 100 per cent government aid. The humanitarian groups, who had put the notion into Schurz's head that the Indians could be rapidly turned into farmers, were still satisfied with results. They were incurable optimists, and as long as they could read hopeful reports (usually false) from the Indian reservations and knew that the government was spending satisfyingly great amounts in carrying out their pet schemes, they were content. They and Secretary Schurz were confident that time would achieve the miracles they were seeking; but, in a free democracy, time and the electorate will wait for no men to execute their plans in undisturbed leisure.

It was election year in 1880 and the Republicans were worried over signs that indicated the danger of their losing their control in Congress. Patronage was needed to strengthen the chances of the party, and the politicians turned to the Indian Service in their search for federal jobs for Republican wheel horses. At Rosebud reservation, John Cook, who knew nothing whatever about either Indians or farming, was made agent. At Cheyenne River Agency, Captain Theodore Schwan, who was excellent at driving his Indians to work in their corn patches, was replaced by Leonard Love, a nonentity but a good party man. At Lower Brûlé and Crow Creek, the able agent, Captain William G. Dougherty, was also displaced, to make room for a political appointee. At Standing Rock Agency, the Roman Catholic church was in control and a Catholic priest had been appointed as agent. For some reason he was left undisturbed. At Pine Ridge, McGillycuddy was also left to finish his five-year term as agent. He had made many enemies by his forthright utterances and outbursts of temper, but he was popular in Dakota and Nebraska, and from a Republican political angle he was a real asset, to be kept and cherished.

These changes of agents put an end for the time being to any hopes that the government's farming plans among the Sioux could show results. At Cheyenne River, the Sioux took off the citizen clothing Captain Schwan had forced them to wear, put on blankets painted their faces, and spent most of their time attending tribal councils and dances. Farming fell off to an alarming degree. Captain Schwan's discipline gave place to a slackness which some observers depicted as a kind of anarchy. Instead of forcing his Sioux to farm (usually with no crops because of drought), Agent Love let them go on a buffalo hunt. Herds of buffalo had miraculously drifted into the western borders of the reservation, and for a season the Cheyenne River Sioux returned joyously to the old free life of hunting. They killed about two thousand buffalo, feasted for weeks on fresh meat, and brought home enough dried meat to keep them for a considerable time. The advocates of support from farming disapproved of Agent Love's conduct, but the Sioux were very happy. They had buffalo meat plus full government rations, with no farm work, and thousands of buffalo robes to trade for goods they desired at the agency stores.

At Lower Brûlé, farming, which Captain Dougherty had started up with such pains and trouble, slumped badly under the new agent. At Rosebud there was no farming, except among the few white squawmen and some mixed bloods. At Standing Rock, the Sioux had always been inclined to grow little patches of corn and vegetables, and under their Roman Catholic agent they continued to do so; but their total in crops was so small that they were still receiving full government rations, with clothing and other needs. At Pine Ridge there was no farming, and McGillycuddy was outraging the Eastern idealists by denouncing their farming scheme as a fraud. He said bluntly that climatic and soil conditions at Pine Ridge made successful farming impossible, and when the Eastern crusaders dared to contradict him, he attacked them vigorously. Haughty and stiff as a poker, he flung in their faces a curt rejection of their plans. "I can confidently venture to state," he reported officially, "that if the experiment were tried of placing 7,000 white people on this land, with seeds, agricultural implements and one year's sustenance, at the end of that time they would die of starvation if they had to depend on their crops for food."[1] The number, seven thousand, was the supposed population of the Sioux at Pine Ridge, and every honest man in Dakota knew that McGillycuddy was stating the true situation. Many families of European whites, drawn to Dakota by the glowing accounts sent broadcast by land boomers, were starving in their bleak shanties in the edge of the Sioux reservation. The Sioux, fat and superior with their free rations and treaty rights, looked down with pity and contempt on these immigrant families. When their agents told them that they must farm, they pointed at these poor white families and shook their heads. No, indeed! They were Sioux and had some pride. Try to live like these miserable white grubbers in a hostile soil? No, indeed!

It seemed that whether they had weak agents or strong ones the Sioux of this period could not be pushed along the road which the Eastern Indian Friends spoke of as progress. Toil, as understood among the whites, was an utterly strange thing to most of the Sioux. They had formerly made their own living by hunting; but the drudgery of steady day-by-day toil in the fields was something alien to their whole outlook on life, and they were neither mentally nor physically fitted for life as farmers. A Sioux who looked big and strong could be outdistanced in a week by a weedy white man, used to steady labor. Besides all this, most of the Sioux were convinced that the government owed them a free living for the lands they had given up, and this attempt to force them to work in order to support their families they regarded as an effort to cheat them out of their treaty rights.

With the enthusiasts in the East demanding that the Sioux work, be speedily educated, and made into citizens, the Oglalas

[1] *Report of the Commissioner of Indian Affairs* (1881), 47.

of Pine Ridge, like the Brûlés of Rosebud were still in the spring of 1880 camped close around the agency, living in idleness on free rations. There is a profane tradition at Pine Ridge that Agent McGillycuddy broke up these Arcadian conditions of life by starting a school. His men began work on a small building. The chiefs, seeing what was going on, stalked into the agent's office and asked what this building was to be. McGillycuddy told them that it was to be a school. And, what was a school? What was done in such a place? The agent explained through his interpreter the purposes and activities of a school. The chiefs, looking depressed and worried, went back to their camps. They summoned councils to discuss this serious situation. News of what the agent was about spread through the camps, and presently the Sioux in a panic took down their tipis and fled. The Kiyuksa band, led by Little Wound, was reported to have gone some forty miles eastward before they felt safe and made camp on Yellow Medicine Root Creek. Other bands established themselves on Porcupine Tail Creek and Wounded Knee, and once a week they all mounted and rode to the agency for their free beef and rations. Nothing would induce them, however, to live near the agency after the school was established.

It is true that McGillycuddy started the first little school at the agency at this time and that many of the Sioux were greatly alarmed over this ominous attempt to deprive their children of the perfect freedom they had always enjoyed. But the true reason for the exodus from the vicinity of the agency was the coming of a special agent—General James R. O'Bierne—who was sent out from Washington to induce the Rosebud and Pine Ridge Indians to scatter out on farming land. At Rosebud, O'Bierne obtained no results, but at Pine Ridge, when he called a council and Red Cloud opposed any action, two chiefs (Man-Afraid-of-His-Horse and Little Wound), who were enemies of Red Cloud, took the opposite view and supported O'Bierne and the agent. Thus the Kiyuksas under Little Wound moved eastward all the way to Yellow Medicine Root Creek, while Man-Afraid-of-His-Horse took his Payabya band to a point on the east bank of Big White Clay Creek, ten miles north of the agency. Even Red Cloud's old ally, Red Dog, moved his camp to Wounded Knee Creek, at a point where the trail from the agency to Little Wound's camp crossed the stream. This was the site of the battle in 1890. The Spleen band (Yellow Bear's band in 1868, now led by White Bird), formed a camp on the east side of Big White Clay, some five miles north of the agency, while the No Flesh band and that group of Wazhazhas still with the Pine Ridge Indians moved to Porcupine Tail Creek.

Red Cloud stubbornly refused to budge, keeping his camp on the bank of Big White Clay Creek, immediately opposite the agency and within sight of Agent McGillycuddy's office. With him remained the Loafer band, led by Red Shirt, whose camp was just south of Red Cloud's. High Bear, another Loafer chief, had a camp two miles or so west of the agency. Thus were created at Pine Ridge early in the year 1880 the two divisions, the nonprogressives and the progressives, titles bestowed on them by the Washington officials and the Eastern idealists. Agent McGillycuddy, more skilled in language, promptly dubbed Red Cloud's nonprogressives the blatherskites, and the old chief—having it explained to him by an interpreter that blatherskites were people who lived in idleness on charity, refusing to do anything toward helping themselves—hated his agent more than ever. Again the chief harangued his people, telling them that the rations, clothing, and other supplies they were receiving were their just due in payment for the lands they had given up. There was no charity connected with the matter. As for working, their generation had been hunters and warriors and could not be expected to learn new ways. It was for the new generation to take up this task of learning to live like whites. While saying this, the chief bitterly opposed education, thus condemning a new generation to share one of the greatest handicaps their fathers suffered under—the inability to communicate freely with the whites and learn their ways.

As was their custom, the Washington officials in their attempt to induce the Sioux to settle on farm lands had employed bait. General O'Bierne told the Pine Ridge people that if a man settled on good land and built a log cabin, he would be given free a door and windows for his house and a cookstove; if he built a two-room cabin, he would also receive a heating stove; and when he went to work on his land, he would receive a plow, wagon, mowing machine, and other equipment free. The squawmen and mixed bloods promptly qualified for all these boons, and many of the fullbloods built cabins and were given doors, windows, and stoves. *Kettle-with-legs*, the Sioux called the cookstoves, and these machines were highly popular, but the people did not find living in log cabins to their liking, and every family had a tipi near their house, in which they lived much of the time.

As he had no faith at all in the possibility of farming successfully at Pine Ridge, Agent McGillycuddy concentrated his efforts on encouraging the Sioux to build log cabins, and within a year he had issued three hundred cookstoves, each representing a one-room cabin built by the Sioux with their own hands. Every stream valley for forty miles around was dotted with cabins—cabins in clusters or long lines strung out along the streams, for the Sioux did not like to live alone, and their new settlements were camps with the cabins not too far apart for the people to have close neighbors on every hand. To control these distant camps, Agent McGillycuddy was planning a school in each district, with a male teacher who would have Indian police stationed in the camp under his orders. Led by the squawmen and mixed bloods, many of the Sioux were starting little gardens or corn patches near their cabins, and the government had distributed a number of cattle, to start the Indians at learning to care for stock. The teacher in each camp was to be in charge of farming and cattle herding. Any Indian who did not care for his garden and cattle and send his children to school regularly was to be dealt with by the Indian police. If recalcitrant, the offender was to be brought to the agency and tried before McGillycuddy's private court, which might sentence him to the temporary loss of his ration ticket or to a sojourn in the guardhouse at hard labor. This system the agent proudly termed Home Rule. In practice it partook more of the nature of martial law, McGillycuddy being a man of violent prejudices and of a wicked temper, which once aroused caused him to press on to extreme action.

Even before Secretary of the Interior Schurz visited Pine Ridge in the fall of 1879, Red Cloud and McGillycuddy had fallen out violently and the chief, backed by many of the tribal leaders, had preferred charges against the agent, demanding his immediate removal. When Schurz was at the agency, a council was held and Red Cloud again demanded the agent's dismissal; but Schurz ignored that. The Secretary, with the idea that the Northern Cheyenne outbreak at Camp Robinson in the winter of 1878–79 might lead to an uprising at Pine Ridge, had put the bold McGillycuddy in charge, and there he was determined to keep him.

After the fighting at Camp Robinson was ended, a few Cheyenne survivors of the slaughter had been officially transferred to Pine Ridge. McGillycuddy, who as a rule preferred warlike Indians, regarding them as more upright, honorable, and honest than reservation Indians, had a strong dislike for these Cheyennes, whom he treated with severity and denounced in his printed reports as worthless people who would not work but sat about their camp mourning for their dead relatives. The Cheyennes had made the mistake of publicly displaying their gratitude to Red Cloud, who had befriended them in their time of trouble, and this did not cause McGillycuddy to like the Cheyennes any better. By this date he had assumed the attitude that anyone who was a friend of Red Cloud's was an enemy of his. His second wife in her recent book, *McGillycuddy Agent*, asserts that he felt sorry for Red Cloud, understood the difficult position the old chief was in, and wished to be his friend;[2] but every word and action of this agent from 1879 to 1886 proves that he regarded it as his first duty at Pine Ridge to break Red Cloud, and this he set to work to accomplish with surprising energy and resourcefulness.

Almost at once on taking charge of the agency he hit on the annual distribution of annuity goods as a lever to be used in upsetting Red Cloud and those chiefs who supported him. There were seven Oglala bands at Pine Ridge, and the custom always had been to divide the annuities (blankets, clothing, bolts of cloth, kettles, axes, knives, and other camp equipment) into seven piles on the ground; the chief of each band then took charge of the

[2] *McGillycuddy Agent.*

goods belonging to his people and distributed them, with the aid of his Indian soldiers or police. Naturally, the chiefs and soldiers showed some favoritism, and McGillycuddy (expecting the common Indians to approve his plan) now proposed that he, as agent, should distribute annuities fairly and equally to the heads of families, ignoring the chiefs. There were no shouts of joy at this handsome offer of freedom and just treatment. The chiefs were very angry and the common Indians just about as angry as the chiefs. No one approved, and McGillycuddy had to give way and watch while Red Cloud in all his glory superintended the distribution of the annuity goods, according to tribal custom.

But V. T. McGillycuddy was a hard man to beat. He kept up his attack on Red Cloud, while the white employees preached the doctrines of freedom from the selfish rule of the chiefs. The result was that in the fall of 1881, instead of seven bands to distribute annuity goods to there were twenty-five bands, many ambitious Indians, who in the past had not dared to wear an eagle feather in front of any real chief, having declared their independence and set up little camps of their own. McGillycuddy was destroying the chiefs by a process of multiplication, and presently instead of seven bands there were sixty-three, and the agent was looking forward to a not distant day when there would be as many bands as men, each head of a family being his own chief. Then, in 1882, he happened to pick up a volume of the United States statutes at large, in which he found in cold print that Red Cloud was right. The law was that annuity goods were to be turned over in original parcels and bales to the recognized chiefs, for distribution among their people. McGillycuddy denounced this law, the Congress that had made the law, and the officials who upheld the law while instructing him and the other Sioux agents to destroy the power of all chiefs.

The drama at Pine Ridge was now heightened by McGillycuddy's formally deposing Red Cloud, declaring him of no more importance than the commonest Indian on the reservation; but almost at once thereafter an order came from Washington for the agent to begin the construction of a fine house, as a gift from the nation to Red Cloud, head chief of the Oglala Sioux. Built in Red

Cloud's camp on the west bank of Big White Clay Creek within sight of McGillycuddy's office windows, this residence was a curious footnote to McGillycuddy's claim that he had deposed Red Cloud, who was no longer a chief. The house was of two stories, whereas the agent's own house was of only one; the Sioux looking at the two dwellings could not escape the conviction that the Great Father considered Red Cloud to be a much bigger man than this boy they had for an agent.[3]

In 1881 (this seems to be the correct date, although McGillycuddy's failure to refer to the matter in any of his reports has caused some doubt whether the event fell in 1880 or 1881), Red Cloud (now fairly and officially deposed) was instructed by the Oglala tribal council as head chief to send out messengers to invite all the Sioux to a great Sun Dance, and McGillycuddy seems to have made no objection, although the Spotted Tail Sun Dance of 1879 had brought forth both official and public condemnation of agents who permitted their Sioux to go off in midsummer to such pagan rites, leaving their little farms neglected for weeks. McGillycuddy, caring nothing for public opinion and little for that of Washington officials, made himself as far as he could the patron of the Sun Dance. He invited army officers to be his guests and attend the ceremonies, some of the officers being eager to take notes and make a permanent record of the curious Sun Dance rites. This was the last big Sun Dance held by Red Cloud's folk. In 1882 the Brûlés of Rosebud put on a great Sun Dance with thirty to forty men making the sacrifice and being tied to the pole for torture, and in 1883 the Rosebud Indians held their last Sun Dance with only three men submitting to torture. That ended the Sun Dance in its old form. During the New Deal administration

[3] The Red Cloud mansion stood on a low ridge about one-half mile from the agent's office, west across the creek. Like Spotted Tail's mansion, it has now disappeared. McGillycuddy's old office is still in use. These mansions for chiefs were another social experiment, the idea being that the chiefs' families would live like white people and thus set an example which the common Indians would quickly follow. In practice none of the wives of the chiefs could learn to take care of a house; there was no free furniture given with the houses and none was bought. After camping out in the bare mansions for a time, even the Sioux were driven out by the accumulations of dirt and smells, and the big houses stood empty.

in the 1930's the Sun Dance was revived by government initiative at Pine Ridge as a part of the new policy of preserving Indian cultural values. Put on by college boys, it was a very ladylike performance—something that would have bewildered the Sioux of the old generations.

McGillycuddy took advantage of the gathering of thousands of Sioux at the Sun Dance to make a display of his new police, having the entire force march to the flats south of the agency where some 15,000 Sioux were encamped in tipis set in a vast circle. Here the police put their ponies into a dead run and swept around the circle of tipis, all whooping and firing their repeating carbines as rapidly as they could work them. The Sioux looked on in silence. They were still united in opposition to the police, and they did not believe McGillycuddy when he stated that the force was necessary to prevent crime. There was no crime at Pine Ridge. The other excuse, that the police were necessary to prevent a possible outbreak, only made the Sioux angry. Who at Pine Ridge was foolish enough to wish to fight the whites?

While McGillycuddy was playing at preventing an uprising among his Oglalas, the agent at Standing Rock was quietly dealing with all of the really dangerous Sioux: the Sitting Bull hostiles. These Indians had found life in the land of the Great White Mother far from congenial. There were no free rations or blankets in Canada, little game for them to hunt and live on, and when they attempted to pick up a living by raiding across the international line, they found the redcoats of the Northwest Mounted Police very hard men to get on with. They simply had no freedom, according to Sioux standards, and before long Sitting Bull began to lose people, hungry and ragged families slipping away from his camp, to cross the line and surrender at Indian agencies in Montana and North Dakota. Some of these former hostiles turned up at Pine Ridge, and McGillycuddy recruited them as stiffening for his already stiff police force. Other agents did not consider it quite nice to hire Sitting Bull bad men to police their friendly agency Indians.

The winter of 1880–81 was that of the big snows and the cold was intense. That winter finished the Sitting Bull die-hards in

Courtesy Pitt Rivers Museum

Red Cloud

Red Cloud in Old Age. Reprinted, by permission, from Hamilton and Hamilton, The Sioux of the Rosebud, *pl. 230.*

Canada. They had already had enough of chronic hunger and going in rags; but that winter was too much, and it was not yet over when groups of starving hostiles began to straggle in at Poplar River (a northern branch of the Fort Peck Indian Agency). By spring the place was thronged, and part of the Sitting Bull people (now fed and warmly clothed at our government's expense) forgot their recent sufferings and their manners. As wild as any Sioux that had ever lived, they knew no method for dealing with white men except that of taking what they desired and fiercely threatening to kill if not given more. They soon created such a condition at Poplar River that troops had to be sent to the scene. Sitting Bull with a small following fled back across the border; but a great majority of his people had had more than enough of Canada. They stayed where they were and surrendered unconditionally to the troops without fighting.

General Nelson A. Miles sent them to Fort Keogh, Montana, and from that point they were brought down the Yellowstone and Missouri rivers in a steamboat to Standing Rock Agency. In July, Sitting Bull with the few followers he still had (157 including women and children) came to Fort Buford, Dakota, from which point his group was taken down to Standing Rock. From here the old chief and his immediate followers were sent down to Fort Randall, as it was considered best to give them a taste of military control.[4]

The Roman Catholic priest, Father Stephan, who had long been agent at Standing Rock, was relieved and went to Washington to head the Roman Catholic mission bureau which was interested in obtaining government subsidies for Catholic mission schools among the Indians, and his place at Standing Rock was taken by James McLaughlin, a Roman Catholic, who had made a fine reputation as agent for the Sioux at Devil's Lake, north of the Missouri River. When McLaughlin reached Standing Rock in 1882, he found the Sitting Bull people camped close to the agency, guarded by troops. He took charge without a trace of parade; the military guard was removed, and he put his wife's relative, Philip

[4] A few of the Sitting Bull Sioux remained in Canada, where their descendants still live.

F. Wells,[5] in the camp, to live with and keep watch on the former hostiles. Sitting Bull and his personal followers were now brought from Fort Randall and added to the camp. The chief was as haughty as in the days of his great successes. He made a list of chiefs in his little group, in the form of colored pictures representing the men, from which it appeared that nearly every man and boy in his camp was a chief. He demanded that all his chiefs should be put at the top of the agency rolls, as they were much more important than any tame agency chiefs. McLaughlin ignored his demands and had his employees plow a small field near the Sitting Bull camp. He then issued hoes and seeds to Sitting Bull and all his big chiefs. That shocked them. Sitting Bull came to the office and explained that his men were new to agency life; they had not yet settled themselves comfortably. It was too soon to think of farming. In a year or two, perhaps. . . . And, was it quite good manners to urge famous chiefs–Sitting Bull, Gall, Crow King, Running Antelope, and Rain-in-the-Face–to bend their backs over hoes? Agent McLaughlin did just that, persistently but with much good humor, and presently the women and even some of the leading heroes in Sitting Bull's camp were planting corn.

The return of the hostiles from Canada created more excitement on the reservation than had been known since the war days of 1876–77. Nothing else was talked of for weeks. A great many Sioux families had relatives in Sitting Bull's camp, and Standing Rock was soon thronged with visitors. McLaughlin issued free rations to 223 Sioux visitors in one week; they all wanted to stay for months, and as more and more came in, it began to look as if all the Sioux would desert their little farms and cattle, to go to Standing Rock and spend half a year with the hostiles. McLaughlin gently shoved the visitors out after a few days, to make room for new arrivals; the other agents, growing strict in the matter of

[5] Philip F. Wells died at Pine Ridge, January 2, 1947. His mother and the mother of Agent McLaughlin's wife were sisters and fullblood Santee Sioux. One sister married Mr. Wells's father, the other married Duncan Graham, a Scottish fur trader from Edinburgh. James McLaughlin was greatly aided in his work as Indian agent by the fact that his wife was half Sioux and spoke the Indian language fluently.

passes, managed to keep most of their Sioux at home. The visitors all wanted to take their relatives in the hostile camp home with them, but orders from Washington forbade any hostiles leaving Standing Rock.

McLaughlin soon reported that it would be safe to break up the hostile camp and send groups back to the agencies where their relatives lived. Agent McGillycuddy was now instructed to send a train of one hundred wagons and a herd of beef cattle across country, four hundred miles, to Standing Rock. There his men gathered the Oglalas and Brûlés in the Sitting Bull camp and brought them home to Pine Ridge, the Brûlés going on eastward to Rosebud.[6] Most of the Oglalas thus returned home belonged to the Crazy Horse camp which in 1877 had fled to Canada, led by Chief Big Road (*Canka Tanka*). This chief now led them back to the Oglala agency and with him was Chief Little Hawk, Crazy Horse's uncle, now a very old man. McGillycuddy talked with the returned hostiles, liked them, and assigned them lands on Wounded Knee Creek, twenty miles from the agency. He gave them fifty wagons and enlisted several of their men for his police force. No other agent imitated his example. They did not like the idea of setting armed hostiles to police friendly agency Sioux; but McGillycuddy stated that these Sitting Bull Indians were less spoiled, more manly, and more reliable than reservation Sioux. He might have added that they were more dangerous, being unused to the ways of the whites, uneasy, suspicious, and given to violent fits of agitation. In a crisis they were liable to give way either to panic fear or to blind rage, and in either case they would probably start shooting. This was what happened in 1890. Without the former hostiles to take the lead, the messiah madness would have been a tame affair.

Unlike the Washington officials and the leaders of the Indian Friends, McGillycuddy seems to have had no faith in the theory

[6] McGillycuddy's report for 1882 shows that he brought 600 Oglalas and 400 Brûlés from Standing Rock. The returned hostiles at Pine Ridge settled to the east of the agency; the Brûlé hostiles formed camps on White River near the mouth of Black Pipe Creek. In May, 1882, 1,000 hostiles were sent from Standing Rock to Cheyenne River Agency and the Rosebud Brûlés have a tradition that they went to Fort Bennett at Cheyenne River and brought many relatives from the hostile camp home with them.

that the Indians would get along better if all chiefs were deprived of power and the tribal organization broken up. He knew the Sioux well enough to understand that it was impossible to deal with these Indians except through chiefs, and from the moment he took charge at Pine Ridge in 1879 he began to consider Young-Man-Afraid-of-His-Horse as a likely successor to Red Cloud. But first, Red Cloud must be deprived of power and perhaps even exiled. Within a year McGillycuddy thought that he had won his battle against the old chief. He deposed Red Cloud. But the Oglalas of Pine Ridge refused to accept his decision and continued to put Red Cloud forward on all official occasions as their head chief and spokesman. He was looked upon as such by the 15,000 Sioux who came to the Sun Dance in 1881, and to clinch the matter, when the officials in Washington got into trouble over the Ponca land affair and needed the aid of the Sioux chiefs to get them out of their difficulties, they forgot all their strong talk about the bad influence of chiefs and the great need to put all the Indians under the control of white agents; they forgot it all, and called on Red Cloud for assistance. If McGillycuddy hoped that the Oglala council would choose his candidate, Young-Man-Afraid-of-His-Horse, to represent them in Washington, he was bitterly disappointed. They chose Red Cloud, and the disappointed agent had to go East as bear leader for the old chief. He had to sit in councils and see Red Cloud treated with every mark of honor and respect, to stand by while important officials thanked this chief for his generous help in the Ponca land affair. These quaint companions—McGillycuddy and Red Cloud—returned to Pine Ridge in the autumn of 1881, and at once fell to fighting again.

McGillycuddy had claimed in print in 1880 that in one year he had saved $50,000 of public money by economizing on rations and other supplies, and in 1881 he claimed another $50,000 in savings. By this he won the hearts of many members of Congress, who were anxious to cut down the cost of feeding and caring for the Sioux. The squawmen at Pine Ridge had read of all this and were discussing the subject when Red Cloud got wind of the matter and asked to have it explained. What the squawmen told him produced an instant outburst of wrath. Red Cloud knew that

Congress had been cutting down the funds for Sioux rations and supplies for some years and he was very bitter about that, believing as he did that the treaty provided a certain amount of food and clothing for the Sioux and that there was no possibility of cutting the funds without robbing his people. And here was this agent claiming that he was saving on rations. Thereby he was pleasing these men in the Great Father's council, and making it harder than ever for the Sioux to get rid of him and obtain an agent that they could trust and honor. Red Cloud now began to accuse McGillycuddy publicly of starving his Indians and of doing it to make friends with the pinch-pennys in Congress. McGillycuddy, becoming annoyed in his turn, said pish and pshaw; that the old fool knew nothing about the ration situation at Pine Ridge, could only count on his fingers, and had not an iota of evidence on which to base a charge of fraud. That was certainly correct, but Red Cloud had friends in both the West and East—men who admired him and men who wished to be his friends simply because McGillycuddy had rubbed them the wrong way and they hated him. And now a cry arose among the Eastern Indian Friends and in the press against the agent, and the *New York Tribune* angrily repeated Red Cloud's charges that McGillycuddy was starving his Sioux.

Red Cloud and his friends among the whites continued their vociferations, while McGillycuddy fumed and snapped out denials. An official investigation held him blameless, but still the angry old chief and his white friends continued to attack him. To this day it is difficult to decide who was in the right. The Sioux at Pine Ridge were certainly not being starved; but did McGillycuddy tell all the truth when he claimed an annual saving of $50,000 on rations and supplies? Congress had already cut the legal treaty ration for the Sioux to about two-thirds of its original amount, and how could this agent save on food without depriving his Sioux of still more of their already reduced rations? At the end of four years McGillycuddy and his supporters were claiming that he had saved $200,000 on rations and other supplies; but was this claim based on anything more solid than a peculiar method of bookkeeping? McGillycuddy was reporting an average number of 7,200 Indians at his agency year after year, but when he was at last removed from office in 1886, his own close friend, Captain James M. Bell, was ordered to make a careful count of the Indians at Pine Ridge, and that census showed 4,873.[7] The claim of McGillycuddy's friends that the Indian Office knew all along that the population figure was padded is strongly vouched for, but what can one think of an Indian agent making public and repeated claims to large savings on rations and supplies when this was made possible simply by estimating for two thousand more Indians than he actually had to feed and clothe? It was no wonder that he could exhibit a saving of $50,000 a year and still feed and care for his Sioux on a handsome standard. Agent J. G. Wright at Rosebud named no names, but he was certainly glancing at McGillycuddy's methods when he reported tartly that no claims for annual savings at his agency were made, as all estimates were carefully prepared and any small surplus left at the end of the year was taken up on the books as a matter of routine. Exactly so. But Wright was Wright, a quiet man who did not seek publicity. McGillycuddy could not live far from the limelight.

Supported by many white persons in the East and by the *New York Tribune* and other great newspapers, Red Cloud's heart was strong, and he redoubled his attacks on McGillycuddy. That agent's principal supporter in Washington, Secretary of the Interior Carl Schurz, had now left office, and the old chief thought the time was ripe to strike a real blow. He obtained advice from some white men, held a tribal council and then dictated a letter to the President himself, demanding that McGillycuddy should be removed, and stating that if this was not done at once, he—Red Cloud—would remove the agent from the reservation. This letter, sent off by mail to Washington, was referred to the Indian Office, which then communicated a copy to McGillycuddy, asking for a report. The report was instantaneous and very loud. McGillycuddy called a council of the Oglala chiefs and demanded Red Cloud's attendance. The old chief declined to obey the

[7] Report of Captain J. M. Bell in *Report of the Commissioner of Indian Affairs* (1886). McGillycuddy was present and aided in the counting, which was done on a fraud-proof system of his own invention.

order. The agent then sent American Horse to bring Red Cloud in, and after a considerable delay the chief arrived, accompanied by about one hundred armed warriors. He sat down in the council room. McGillycuddy shouted at him to stand up, but Red Cloud ignored him. A violent scene ensued, Red Cloud finally getting to his feet while McGillycuddy shook the offensive letter in his face, shouting denunciations at him. He wound up by announcing that Red Cloud was deposed and of no further importance. American Horse, he said, was from this day the chief of the Smoke People (Red Cloud's division of the tribe), while Young-Man-Afraid-of-His-Horse was to be the new head of the Bear People.[8]

Our knowledge of this affair is faulty. We are not told, for example, why Red Cloud, backed by one hundred armed warriors, gave in simply because McGillycuddy shouted at him. One suspects that on this occasion as on several others the agent solemnly warned all the assembled chiefs that if they did not desert Red Cloud and support him, the officials in Washington would send troops. Frightened by such a threat, the Sioux always deserted Red Cloud. McGillycuddy's friends claimed that this flare-up was entirely the work of the vindictive old chief, but the whole matter smacks of action by the Oglala tribal council, with Red Cloud merely acting as spokesman for the tribe. How otherwise are we to explain McGillycuddy's action in deposing Little Wound along with Red Cloud? Little Wound was Red Cloud's enemy, a strong supporter usually of McGillycuddy; but on this occasion if the tribal council decided to demand the agent's removal, Little Wound had to acquiesce. So old Little Wound was deposed, to make room for Young-Man-Afraid-of-His-Horse, who was McGillycuddy's favorite.

Blaming Red Cloud and his blatherskite followers for this attempt to remove him, McGillycuddy reported officially that Red Cloud had lost all influence among the Oglalas because of his heavy drinking. Liquor, wrote the agent, "is not unwelcome

[8] *South Dakota Historical Collections*, XIV, 523. This is Doan Robinson's account, giving no exact dates. The official reports have not been uncovered; but the agent and Washington officials mention the Red Cloud affair in 1881, and some of them make the absurd claim that war was narrowly averted by McGillycuddy's bold conduct.

even to Red Cloud (whom an editorial in one of the philanthropical journals East recently and very gushingly termed the grand old chieftain), for excess use of the fluid which exhilarates and at the same time intoxicates has had much to do with eliminating what grandeur formerly existed in this Indian, and has resulted in his downfall among his people."[9]

Was this merely a fresh display of McGillycuddy spleen, or did this agent have evidence on which to base these charges? He included Spotted Tail in this indictment for drunkenness; yet in all the contemporary records there is no hint that either Red Cloud or Spotted Tail drank to excess. Then there is the other assertion, that Red Cloud had lost all influence among the Oglalas, and here we have McGillycuddy's own reports for 1882, 1883, and 1884 to refute what he wrote in 1881; for in these years he reports Red Cloud as the accepted head chief at Pine Ridge. Moreover, after Spotted Tail was murdered, all the Sioux sent delegations of chiefs to Pine Ridge, to ask Red Cloud to be the head of the entire nation.[10]

In the summer of 1882, Red Cloud and McGillycuddy put on a new and uproarious performance, and this time Red Cloud played the incredible part of labor agitator, while McGillycuddy was most unwillingly forced to accept the role of villainous employer. There was now a railroad running through northern Nebraska, along the southern borders of Rosebud and Pine Ridge, then northward to the Black Hills, and the town of Thacher on this line was only 130 miles from Pine Ridge, whereas Rosebud Landing on the Missouri, from which point all supplies were being hauled in wagons, was 190 miles away. The saving of sixty miles of haul meant a huge reduction in the cost of wagon freight to the agency, and the Indian Office ordered McGillycuddy to make the change. On hearing of this, the Sioux who were hauling freight from Rosebud Landing became violently agitated; councils were held, and Red Cloud was put forward to play the labor czar and demand that the government continue to do things in the old and expensive way, so that the wages of the Indian teamsters would

[9] McGillycuddy, in *Report of the Commissioner of Indian Affairs* (1881) 45.
[10] *Dawes Report*, 48 Cong., 1 sess., *Sen Report No. 283*, 23.

not be reduced. Red Cloud, still sore from the pommeling McGillycuddy had given him in 1881, willingly undertook to play the hero role in this new melodrama. He went to the office and demanded an interview. Steel struck flint, and the sparks spurted. Red Cloud demanded that the order for the change should be set aside at once, and McGillycuddy angrily told him to get out, that he was busy and had no time for nonsense. Red Cloud got out, with the whiz of a rocket. He called a tribal council, and the chiefs decided that there should be no change in the wagon road. They put the tribal soldiers in control of the situation with orders to deal severely with any teamster who obeyed the agent and attempted to haul loads from the railroad. Any such offender would be quirted, have his harness cut up and his wagon broken. If he persisted in his wrongheadedness, the Indian soldiers would give a second and more severe treatment. The teamsters were delighted. Now their union was in a very strong position with Red Cloud, the council, and the tribal soldier-police all supporting them.

It was McGillycuddy's move, and he moved instantly, vociferating as he sped along. This was an Indian insurrection, and the wicked Red Cloud had instigated it. The wrathful agent at once resorted to that ancient appeal to the Sioux "supreme court" —the stomach—ordering Red Cloud and all his principal supporters cut off sugar, coffee, and bacon. They were to have just plain beef and flour. Such disastrous orders from their agent had struck the Sioux in their midriffs many times in past years; but the cutting off of their rations always came to them, as it did now, as a unique and unheard-of thing, an incredible act of tyranny. To make sure that all opposition should end, McGillycuddy also ordered his police to disband any further tribal councils.

Reaching into the air for a weapon, Red Cloud drew down the strangest one yet—a legal weapon having to do with neutral territorial rights, and the best of it was that McGillycuddy himself had made this weapon and laid it within easy reach. On coming to Pine Ridge McGillycuddy had at once become involved in a violent quarrel with the Nebraska citizens who had established ranches hardly more than a mile south of Pine Ridge Agency. All of these men were trading with the Indians, and some of them were selling liquor. To put a stop to this, the resourceful McGillycuddy proposed to the Indian Office that a neutral strip five miles wide and ten miles long should be established in Nebraska, running along the south line of the reservation, and that all Nebraska citizens should be compelled to move out of this territory. He kept after the Washington officials, writing many urgent requests, and in the end a Presidential order was issued establishing the desired neutral strip into which neither the Nebraska men nor the Pine Ridge agent or his police might go; but the Sioux were permitted to camp in this tract and to cut wood there.

Red Cloud now skipped down into this neutral belt, taking his friends with him; and there he called a new council, leaving McGillycuddy and his police to rage as loudly as they chose. They could not touch him. Taking full advantage of the agent's order cutting off sugar, coffee, and bacon rations from his followers, the old chief went to visit his Nebraska friends and told them very earnestly that McGillycuddy was starving the Sioux again, and much as the Sioux would dislike to injure their friends in Nebraska, if this went on much longer, the Indians might have to come across the line and take cattle wherever found to feed their families. A rumor of impending Indian war spread along the border and excited editors began to denounce McGillycuddy in print. He was a menace to peace; he must be gotten rid of at once. Meantime the Sioux were flocking into the neutral strip, and the council dictated a message to be sent to Washington demanding that McGillycuddy should be removed and stating that if this were not done within sixty days, the Sioux would remove him themselves. This is Herbert Welsh's version, in the Indian Rights Association Report for 1883. McGillycuddy (and his wife in her recent book) imply that Red Cloud and a few blatherskites made all the trouble, while the council and "progressive Indians" were on the agent's side. In truth the council and the whole tribe were on Red Cloud's side until covert threats of bringing in the cavalry frightened them into deserting the old chief.[11]

[11] McGillycuddy glanced spleenfully at these events in his report for 1883. He said that the trouble was caused by Red Cloud, "the so-called chief of the

All in all, this full-dress performance at Pine Ridge in August, 1882, was the most wonderful yet. Nothing to compare with it had ever been seen on the Sioux reservation, and as an example of what McGillycuddy's friends termed his perfect control over his Indians, it was simply prodigious. There was now a telegraph office at Pine Ridge and the agent, while posing as a strong and unperturbed man, in complete control of the situation, was sending frequent telegrams that made the hair of the officials in Washington stand on end with alarm. He had mobilized for war, his armed police were galloping about on war ponies, and he was issuing arms to Indians he called the Friendlies, as if all his Oglalas were not that. Most of his Friendlies were men who disliked Red Cloud and would be glad to take a shot at him under circumstances in which they could pose as heroes serving the government loyally. The papers in the National Archives show that McGillycuddy issued arms to selected groups and that he closed the agency buildings and put the place under a kind of state of siege. He also dismissed one of the agency traders on the grounds that he had sided with Red Cloud; in fact, the trader was sympathetic toward the Oglala tribe, so he was labeled disloyal and ordered to leave.

Through all this excitement it was plain enough that no one at Pine Ridge wanted war. The Indians knew that if they started shooting they were ruined. They would stick to Red Cloud up to the moment when a threat was made to bring in troops, which meant big trouble ahead. This point was now reached. McGillycuddy finished his telegraphing and called a council, which was attended by the group he termed the Friendlies. He talked to the chiefs and headmen about the serious trouble Red Cloud was leading them toward. Troops were under arms at Camp Robinson in Nebraska. He, their agent, had long ago promised them that he would never call for troops, and he would keep that pledge; but the officials in Washington had made no pledge, and they were very much upset by Red Cloud's doings. If the chiefs and head-

Oglalas" who "broke out in opposition to the progressive Indians." Here he keeps up the pretense that Red Cloud was not a chief. On the Sioux reservation "progressives" meant Indians who worked, and about the only Pine Ridge Indians who were working in 1882 were the teamsters engaged in hauling freight. They were on Red Cloud's side to a man when this trouble started.

men would promise their support, this matter could be dealt with without the calling in of the cavalry.

The chiefs, who had no desire whatever for a visit from the cavalry, deliberated and then informed McGillycuddy that they would support any action he thought necessary. These chiefs, some of whom were Red Cloud's hereditary enemies, now made their arrangements, while the agent made his, and when all was ready Red Cloud was summoned to the agency. He declined to come. Little Wound (whose father, Chief Bull Bear, Red Cloud had killed on the Chugwater near Fort Laramie in 1841) was the principal chief among the Oglalas whom McGillycuddy called the Friendlies. Appealed to, Little Wound sent a runner from the agency to his own camp on Yellow Medicine Root Creek, summoning all the Kiyuksa warriors with arms to Pine Ridge. McGillycuddy warned his police that an attempt to arrest Red Cloud was to be made. One of the policemen resigned at once; the rest were loyal to the agent's salt. Young-Man-Afraid-of-His-Horse now informed McGillycuddy that he and Little Wound and the other chiefs were going to summon Red Cloud in the name of the Oglala tribe. Such a summons no Oglala dared to ignore. Red Cloud came, accompanied by a large group of armed supporters. The arrangements were neat. Red Cloud was seated in the council room with his principal enemy, Little Wound, close beside him, and Little Wound had a cocked Winchester under his blanket, the muzzle protruding and aimed straight at Red Cloud's midriff, which it almost touched.

McGillycuddy now made a violent speech, attacking Red Cloud. Flourishing a yellow telegraph form, he shouted that here was the government's answer to Red Cloud's defiance—an order that this chief should be arrested, if that were necessary to preserve peace. Mrs. McGillycuddy in her narrative states that when the telegram was explained by the interpreter, Red Cloud leapt up with a knife in his hand and rushed at the agent, but Little Wound jumped up and stopped him.[12] McGillycuddy then deposed Red Cloud (was this the third or fourth time?) and ordered his arrest.

[12] *McGillycuddy Agent*, 196. I find no other reference to Red Cloud drawing a knife on his agent. Mrs. McGillycuddy in this book gives a partisan account

That sojourn in the guardhouse at Pine Ridge hurt Red Cloud terribly. He was a proud man, and to be thus treated by "that boy in the office," whom he despised, was too hard a fate. What were the Oglalas coming to when the head chief's own nephew would obey a white man and put his family and tribal leader in jail? His nephew, Sword, chief of police, had done just that.[13] For the moment the old man's spirit was broken. He gave a promise of better behavior and was released. Once out of the hateful guardhouse he streaked it for Nebraska, shaking the dust of the reservation from his moccasins. His intentions were of the best, but the devil was baiting a new trap for him. At the little town of Thacher he found awaiting him at the railway station a fine team with wonderful shiny harness and a new wagon painted in gaudy colors. This, seemingly, was a gift from the government (as a reward for his friendly services in 1879–80). Indeed, it was a new proof of what an important man he was and how much the officials in Washington valued him. Disdaining to soil his own hands, the chief summoned a white man, ordered him to harness the team and drive him back to the reservation. In regal style he was driven into one Sioux camp after another; and in each camp he stood up in his fine new wagon and harangued the people, telling them that this expensive equipment had been sent to him by the Great Father as a reward for his services in protecting the Oglalas from that bad-tempered boy they had for a father.

Thus Red Cloud had the last word; that is, the last word in August; but far, far from the last word in this year 1882. These were wonderful times, happy days for the Sioux, who had been relieved of the boredom of staying at home and caring for their gardens and cattle because Red Cloud and their father the agent could not agree about anything. There were constant councils

of all the events, making her husband the hero. She is far from reliable. She states that immediately after this council her husband took the train at Valentine, Nebraska, and went East. Valentine was not in existence at this time and the journey she records was made two years later in 1884.

[13] *Miwakon Yuhala*, He-Has-a-Sword, usually called George Sword, son of Brave Bear of Red Cloud's own band, the Bad Faces. Some of the Sioux today say there were two brothers called Sword and that the elder was made a shirt wearer or war chief on Powder River during the fighting of 1865–68. They call George Sword the younger brother.

and excitements, the rush of mounted messengers to the camps to warn the men to get ready to support either the chief or the agent. The big performance in August had hardly ended when in September more excitement and happiness was produced by Inspector W. J. Pollock, who came to look into the causes of the August fracas. He called a big council at the agency and visited the camps for councils with each band. Red Cloud, whom McGillycuddy had leveled with the dust in August, was up again, exchanging blow for blow with the agent. The old chief's followers, no longer with the smell of big trouble in their nostrils, were backing him loyally once more.

McGillycuddy had done his best to make Inspector Pollock his enemy, hinting that Pollock was dishonest and a secret member of the mysterious Indian Ring that was alleged to be robbing both the government and the Sioux. Mrs. McGillycuddy in her book states that Pollock during his investigation sided with Red Cloud against her husband; but Pollock's report, dated September 15, 1882, is in the National Archives, and in it Pollock placed all the blame for the August troubles at Red Cloud's door. No other decision was possible. Every weapon Red Cloud had employed was unethical, but how beautifully effective some of them had been! The agent's pose of calm superiority had been cut to ribbons, leaving him for the moment hot, angry, and hoarse from shouting. The absurd and easily refuted charge Red Cloud had hurled at him, that he was starving his Sioux, played havoc with McGillycuddy's reputation. Not only the Nebraska newspapers but some of the most influential ones in the East had taken it up, picturing McGillycuddy as a feudal lord, living in a palace and (while his Sioux victims starved) sitting down to a board groaning with the weight of the finest meats, fish, and imported luxuries. They told of the noble cellar, stocked with imported wines and liquors; they detailed his corruption of inspectors, sent to Pine Ridge to look into affairs, whom he won body and soul by giving banquets in their honor and serving them with vintage champagnes. There was almost no foundation for these handsome accounts of McGillycuddy's home and habits; but after scanning these newspapers, he erupted like a volcano, denouncing Red

Cloud, the blatherskites, the squawmen and other whites, on and off the reservation, who supported Red Cloud and put ideas into his wicked old head; he excoriated the venal press that spread false reports about him; and (forgetting in his wrath their position as Sacred Cows) he blistered the "Men of Good Will"—the "Friends of the Indians," the benevolent brethren of the Eastern philanthropic societies, who now had a regular pressure-group office in Washington from which an alarm had been sounded—a false alarm as McGillycuddy opined, for it was an announcement smacking strongly of support for Red Cloud. Forgetting his manners, the angry agent dubbed these kindly men and women sentimentalists and cranks.

He was so sure of his own rightness; but was he in any degree right? The Supreme Court was about to decide, in the Crow Dog case, that the Sioux treaty was the sole law on the reservation, and there was nothing in that treaty to authorize an agent to usurp the powers of the Oglala tribal council by deposing a chief. If Red Cloud's conduct was unethical, the conduct of the agent and his superiors in Washington was illegal. The Washington officials had no authority for ordering McGillycuddy to put Red Cloud in jail, and their pretense that this illegal act was necessary to preserve peace was a sham. They had just one prime duty to perform, under the treaty, and that was to give the Oglalas of Pine Ridge an agent with whom they could get on quietly. There had been quiet before McGillycuddy came; there was quiet after he left. If there were threats of armed conflict during the whole period of his rule at Pine Ridge, he was himself the principal manufacturer of those threats.

Autumn came, and the Newton Edmunds land commission arrived at Pine Ridge, to hold councils and to induce the Sioux to sign away a large portion of their reservation. The commissioners found a curious situation. At Rosebud they had been warned not to accept McGillycuddy's hospitality, as there was a violent three-cornered quarrel going on at Pine Ridge, where Red Cloud, having been deposed and squashed in August, was again recognized as head chief by the tribe, while McGillycuddy was trying anew to oust him and to put Young-Man-Afraid-of-

His-Horse in as head chief. This younger chief and his supporters were aiding the agent in fighting Red Cloud; the tribe was in the main on Red Cloud's side, and if the land commission accepted the agent's hospitality, the Indians would regard them as his friends and would refuse to have any dealings with them. Thus the land agreement would be lost.

The Edmunds commission took a chance and became the agent's house guests. They called a council, and the Sioux promptly put Red Cloud forward to speak their will. He rejected the land agreement in a spirited speech. McGillycuddy then made an even more spirited speech, which was translated by an interpreter to the listening chiefs and headmen. It made them sit up. He told them to go out and consult among themselves, and that it was his hope (here he looked at them very hard) that they would find on second thought that there was reason for them to change their minds. The Sioux easily found the reason. Here was another smell of big trouble, produced by this agent of theirs, and they promptly shied away from Red Cloud's leadership. They decided to sign the land agreement, because McGillycuddy seemed to mean that trouble would come if they did not. Red Cloud, determined to protect his position as head chief, rushed to the table and was the first man to have his name set down. If he had not done that, some other chief—perhaps Young-Man-Afraid-of-His-Horse—would have signed first, and then many Indians would have regarded that first signer as head chief. Red Cloud pocketed his pride to save his official face and signed a land agreement which he did not approve of, and within a month it became apparent that the agreement was a swindle. The old chief promptly shifted his ground and started an anti-land-agreement campaign, leaving McGillycuddy in the uncomfortable position of being for an agreement which all the Sioux now believed was a barefaced attempt of white men to steal their land.

By this time many men were expressing the view that there would be no quiet at Pine Ridge as long as Red Cloud and McGillycuddy dwelt there together. There were proposals of confining Red Cloud in a fortress on the East Coast; but how could that be done, when he was guilty of nothing more than endlessly squabbling with an agent? Another proposal was to remove McGillycuddy, but the Washington officials could not bear the thought of losing this agent's valuable services—services that consisted mainly in his ability to arouse Red Cloud's wrath and to fight him when he was aroused, to fight but seemingly never to conquer.

During the summer and autumn of 1882, McGillycuddy was not only engaged in the struggle with Red Cloud; he was also involved in a quarrel with the Northern Cheyennes who had been sent to Pine Ridge. These Cheyenne troubles seem to belong to the early autumn, to the time immediately following Red Cloud's arrest and confinement in the guardhouse. The Cheyennes had been brought back from exile in Indian Territory and were being temporarily kept with the Sioux at Pine Ridge. McGillycuddy had a strong dislike for these Indians, and the fact that the Cheyennes looked on Red Cloud as a great chief who was their protector did not help matters.

The Cheyennes wished to camp with Red Cloud, whose camp was immediately west of the agency, but McGillycuddy forced them to encamp on Wolf Creek east of the agency and employed his police to keep them away from Red Cloud's camp. He pressed this fresh quarrel with his usual vigor, and presently the Cheyennes (taking a leaf from Red Cloud's recent criminal record) struck their tipis in the dead of night and moved camp down into the neutral strip along the Nebraska border. Here the agent's police could not follow them; but McGillycuddy had a trump card which he promptly played, ordering the Cheyenne rations cut off until they should obey his orders and return to their camp on Wolf Creek.

The Cheyennes were soon hungry enough to return to the agency, but they came with 150 armed warriors, demanding full rations for their families at once, McGillycuddy met them at the head of his 50 police armed with repeating carbines and shouted them down. Sullenly the Cheyennes went back to their Wolf Creek camp, but on September 23, 200 of them led by Chief Black Wolf slipped away from camp, determined to make their way to their kinsmen on the Tongue River reservation in Montana.

The agent called out his whole police force, issued ammunition, and was on the point of starting after the fugitives when he received telegraphic orders from the Indian Office to desist. It was just as well that some men in authority at times had a sense of proportion and were unwilling to bring on a bloody fight because a few Indians had left a reservation without a formal pass.[14]

It seems incredible, but it is a matter of official record that after his performances in 1882, Red Cloud, the deposed nobody, was invited to Washington during the ensuing winter, where he was greeted as a great chief and an honored guest. He was taken on a tour of inspection of Hampton Institute in Virginia and Carlisle School in Pennsylvania, where large numbers of Indian boys and girls were being trained. The chief was in his mellowest mood, courteous and most friendly, and he made speeches on education to admiring audiences of white gentlemen and ladies, who were delighted to find that the noble chieftain agreed with their view that of all the boons that civilization held out to the red race, education was the greatest. Agent McGillycuddy, fizzing with wrath, wondered if all high-minded folk were born fools. There, in plain view from his office windows, was Red Cloud's camp across Big White Clay Creek, and in that camp were over seventy children of school age, not one of whom was in the agency school because of Red Cloud's hatred of schools. Two boys had gone to Carlisle, but they were mixed bloods whose parents and let them go despite the chief's opposition.

Red Cloud's attitude toward education was first expressed in 1870, when a school at his agency was proposed. He stated that his people still had buffalo and it was too soon to speak of schools. This seemingly cryptic remark simply meant that the chief regarded schools as places in which children whose parents could not feed and clothe them were cared for, apparently by white persons with soft hearts (and heads); but the Sioux still had buffalo meat for their children and buffalo robes to give to traders for cloth with which children's clothing could be made, so schools

[14] These Cheyenne troubles of 1882 are recorded in the *Northwest Nebraska News*, June 30, 1885. The editor had recently visited Pine Ridge and obtained the story from eyewitnesses.

were not yet needed. His people were not paupers. By 1882 the chief's views about schools had developed somewhat. He no longer regarded the white men and women who ran these queer places as weak minded, for he had learned that Indian schools all required much hard work from pupils: gardening and work with all kinds of tools, cooking, bedmaking, scrubbing, and—hardest of all (here he was right)—work with books. He was convinced that the people who ran the schools collected money from someone for all the work the children did, and pocketed it. Thus while Red Cloud's friends in the East imagined that he favored education, just because he made some speeches to that effect, the old man in fact regarded schools as places at which Indian children were enslaved, to work for the enrichment of white men and women. McGillycuddy, who knew Red Cloud's real views on education, smiled wryly and went on establishing new schools. He now had six out in the Indian camps, one of them forty miles from the agency.

The year 1883 was one of great excitement among the Sioux, because of the exposure of the true nature of the Newton Edmunds land agreement of the previous year. Most of the Indians now regarded that transaction as an attempt to rob the Sioux of half their land. Red Cloud, deposed and sent to jail by his agent in 1882, was again the accepted leader of his people, and McGillycuddy, playing a very minor role, looked on helplessly while even those Sioux who generally sided with him and against Red Cloud followed the old chief's lead in the fight to save their lands. The agent was paying heavily for his ill-judged use of pressure to induce his Indians to sign the agreement, and for once he had no justification further than a particularist plea that the agreement would have benefited the Pine Ridge Oglalas—at the expense of the Rosebud Brûlés. Indeed, regarding it as imperatively necessary to win the support of the Oglalas, the men who drew up this land agreement had taken a strip of territory from Rosebud and offered it to the Pine Ridge Sioux. McGillycuddy's argument was that as agent for Pine Ridge it was his duty to act solely in the interest of the Pine Ridge Sioux. If the Sioux at other agencies were robbed of their land, that was their agents' affair, not his.

This excitement over the land question was kept up all through 1883 and on into the following year. Dr. T. A. Bland, a noted Indian Friend and editor of the *Council Fire*, published in Washington, was very active in the campaign against the land agreement. Unlike most of the humanitarians who were interesting themselves in Indian welfare, he was opposed to any attempt to hustle the Indians along the road toward civilization. He advocated letting these people take their own pace. He fought against the policy of breaking the chiefs and destroying the tribal organizations in order to place the Indians completely under the power of white agents. He protested against the carrying off of Indian boys and girls to distant boarding schools in the East, which were little better than penal institutions, and he was strongly against the new theory of the Eastern Indian policy makers, which held that most of the Indian land should be sold, as Indians would never settle down and farm as long as they had great tracts of wild land to roam about in. Dr. Bland, in short, held very similar views to those of Red Cloud, and the two men were good friends. They corresponded, Bland's letters being interpreted to the chief, who then had replies prepared by one of the Pine Ridge squawmen.

Dr. Bland was a former minister. In Washington he was highly regarded by many persons, and the worst that could be said against him was that his views did not agree with those held by a majority of the brethren who called themselves Indian Friends. McGillycuddy, who had been publicly criticized by Bland, had retorted hotly that the man was a sentimentalist, a crank, and a good deal of a rascal. Bland replied in more moderate terms, and there it might have rested, but in 1884 this crusader for Indian rights decided that the Sioux land agreement crisis had made it necessary for him to visit his friend Red Cloud, and—apparently quite unaware that he was about to play the part of Daniel in the lions' den—he packed his carpetbag and set out for Pine Ridge.

Dr. Bland left the railroad in northern Nebraska and came up by spring wagon along a road which was decorated on either side by notice boards all bearing in large letters the signature *McGillycuddy, Agent*. Clearly this was enemy country, but Bland regarded the threatening signs calmly. He was rather an important man in Washington, and as a retired minister he felt that the sacredness of his black clothes was a sure protection. Besides, he had a personal letter from the Secretary of the Interior to this agent requesting every courtesy and consideration for the bearer. Therefore the signboards instructing all visitors to report at once to the agent's office on reaching Pine Ridge did not impress him greatly.

Met by a large number of Indians and squawmen, Bland drove into the agency about noon, followed by several wagons in which Red Cloud and other chiefs and the squawmen were riding. Having escorted their distinguished friend to the mess hall, the Indians and squawmen went to their camps to eat, while Dr. Bland ordered a meal and prepared to rest after his fatiguing journey. Meanwhile, McGillycuddy's police, always on the alert for the appearance of suspicious characters, had observed the arrival of this black-coated man and his caravan which included Red Cloud and his leading blatherskites. Reporting at the agent's office, Chief of Police Sword remarked on the suspiciousness of all this; why had this stranger ignored the notice-board warnings and failed to report at once and give an account of himself? McGillycuddy sent Sword and twenty policemen to the mess hall, where they forced Bland to leave the meal he had just sat down to and escorted him with military pomp to the agent's office. Bland stalked in, bristling with anger, but he had hardly started to demand the reason for this outrage when McGillycuddy cut him short with a curt statement that he must leave the reservation at once. His purpose in coming, said the agent, was known. He was here to assist and advise Red Cloud in his disobedience. Dr. Bland flourished the letter from the Secretary of the Interior, but McGillycuddy brushed it aside. He said that he had a copy, and that although the letter requested every consideration for the bearer, it stated explicitly that the bearer would receive such courtesy only if he conformed to the agency rules and regulations. This he was not doing, this he had come prepared not to do, and his conduct would not be tolerated for one moment.

Dr. Bland's visit at Pine Ridge lasted less than one hour and ended in his being hustled back into his spring wagon and escorted down to the Nebraska line, this time by twenty armed and mounted Indian policemen. Once started on such a campaign, McGillycuddy knew no restraint. He summoned the squawmen who had escorted Bland to the agency; he gave them a scathing lecture and then banished them, some to their distant ranches, others being ordered to leave the reservation. He then had fresh notice boards set up warning all squawmen and mixed bloods not to harbor or give aid inside the reservation limits to three undesirable characters whom he listed as R. C. Clifford, Squawman; T. A. Bland, Philanthropist; and Tod Randall, Squawman. This sandwiching of the worthy doctor between two squawmen of allegedly unsavory character was a typical McGillycuddy touch.[15]

The raging Dr. Bland halted at a ranchhouse in Nebraska and made the place his headquarters. Red Cloud and his followers came there to council with him, as did also a host of squawmen, mixed bloods, and Nebraska frontiersmen, all of whom had at one time or another suffered from McGillycuddy's rods and scorpions. Their voices raised in angry debate reached as far as the agent's office at Pine Ridge; indeed the whole reservation was seething with excitement. Presently, as Mrs. Gillycuddy reports, the malcontents issued an ultimatum. McGillycuddy must be removed within three days, and if the officials did not act, the Sioux would themselves remove or kill him. There is no official confirmation of this statement.

McGillycuddy then played his trump card, the same one he had played in each crisis since 1879. He called a council of the chiefs and solemnly warned them that if they did not abandon Red Cloud's cause at once, they would find themselves in most serious trouble. He made his usual statement, that having pledged his word in 1879 never to ask for troops, he would keep that pledge, but there were higher officials who had made no pledge,

[15] In *McGillycuddy Agent*, Mrs. McGillycuddy gleefully records this meeting between her husband and Dr. Bland, giving no date. A letter from McGillycuddy to the commanding officer at Camp Robinson, dated August 5, 1884, shows that Bland had been removed from the reservation shortly before that date. The letter is in the National Archives.

and already the troops at Camp Robinson had been warned and were preparing to march.

The chiefs instantly deserted Red Cloud. They wanted no soldiers at Pine Ridge. Dr. Bland went back to Washington, and Red Cloud returned to his camp across the creek from McGillycuddy's office. The usual official investigation followed, a special inspector (General McNeil) coming out from Washington. McGillycuddy called a council, and Red Cloud (who had been deposed three or four times but was still looked on by his own people as head chief) opened the ceremonies by denouncing his agent as a tyrant, thief, and general nuisance. McGillycuddy sprang to his feet shouting that Red Cloud was a liar, also a squaw. Chiefs sprang up on every hand, shouting accusations, some aimed at the agent, others at the head chief. General McNeil sat rigid with amazement as the mighty brawl roared on and on. Presently the agent went out, abandoning McNeil in the wolf den. Much later McNeil escaped, his brow beaded with perspiration. He sought McGillycuddy and told him that although he was a veteran of many Indian councils, this was the damnedest performance he had ever witnessed. He asked if it was the usual thing at Pine Ridge, and the agent admitted grimly that it was. Just a fair sample of what his admirers termed his perfect control over his Sioux.

McNeil spent three weeks at Pine Ridge, carefully examining into the state of affairs, then wrote a report clearing McGillycuddy of all blame. No other view was possible. Red Cloud was using every weapon that came within his reach in this long struggle, but most of his charges against McGillycuddy were as poorly supported as his main contention, that this agent was stealing from both the Indians and the government. The old man may have believed that, but no one who had examined the agent's immaculate accounts would give the charge a second thought. Indeed, Red Cloud had only one just charge against McGillycuddy and that was incompatability. What he was seeking was a divorce, but the officials in Washington—all on McGillycuddy's side—had no intention of giving him one. If McGillycuddy was a hair shirt, that was exactly what Red Cloud needed to keep him in a humble mood.

Red Cloud, feeling anything but humble, looked about for a new weapon, and of all the unlikely objects he might have been expected to seize and hurl at McGillycuddy's head, he picked out a boarding school! This school was the agent's own favorite project: a fine new building to house one hundred Oglala boys and girls, who were to be trained industrially right there at home in Pine Ridge. McGillycuddy had labored hard to get the initial $20,000 appropriation out of a parsimonious Congress; he had supervised the planning and construction of the school, and in the summer of 1884 he had it completed and equipped. He then made a special journey to Omaha in Nebraska to meet the staff of female teachers and attendants he was importing from New England.[16]

When he returned to Pine Ridge with this female faculty, he called a council, introducing the ladies to the chiefs, and then harangued his Sioux on the wonderful opportunities that were being offered to them and their children by a kindly government. Red Cloud instantly attempted to break up the meeting with a cry of no schools; but for once the Oglalas would not hear him. To have their boys and girls live in this palace (they called it *Owayawa Tonka*, Big School), fed and clothed free, waited on and cared for by real white ladies from the East, was an offer too tempting to be resisted. McGillycuddy was swamped with applications, the chiefs and headmen leading in insisting that their own boys and girls should be admitted to this paradise. Even Red Cloud went back on his principles and had his small daughter enrolled.

When the opening day came, the agency was jammed by excited groups of Sioux who had ridden in from distant camps to witness the ceremonies. But there were no ceremonies. The anxious parents brought their boys and girls to the imposing entrance of *Owayawa Tonka*, the door opened, a haughty white female snatched in the child and instantly shut the door in the faces of the

[16] Here we have a date in the National Archives—McGillycuddy signed the contracts for his school staff July 1, 1884, and brought the women to Pine Ridge early in the autumn.

anguished parents. All the shades on the many big windows had been drawn down and the eagerly watching crowd of Sioux could see nothing. From the interior of the big building muffled and mysterious sounds were heard. Filled with misgiving, the Sioux discussed the probabilities of what was transpiring behind those lowered blinds. The wind presently brought tidings. It blew aside a blind on a lower window, and an Oglala man who was staring at that particular window caught a hurried view of dire proceedings. He beheld a white woman holding an agitated Sioux boy down on a chair while another woman bent over him, lifting his scalplock with one hand while she poised a big pair of bright shears in the other hand. The warrior let out an eerie yell. "*They are cutting the boys' hair!*"

Pandemonium broke out, and Sioux parents stormed the door, most of them getting inside in time to save their sons' sacred scalplocks; but others, too late, came out with shorn lambs, wailing with grief. Other secrets of the big tipi were disclosed. Oglala girls, naked and tearful, had been found seated in tubs of hot water with grim-faced white women at work, scrubbing them with soap and brushes. Agent McGillycuddy circulated through the roaring crowd shouting that it was all right, everything was perfectly all right—but the Sioux parents did not agree with him. Red Cloud rescued his daughter from the dreadful building and bore her away, nearly all the other parents following his example.

McGillycuddy called the usual council and tried to explain to the glum chiefs the need for cleanliness in a boarding school, but his listeners were hard to convince. In the matter of haircutting they were adamant in their opposition. The scalplock was the Sioux badge of honor, a boy deprived of that lock would never amount to anything and would soon become a social outcast. The agent kept at them. The school must be filled again. Some concessions were made to Sioux prejudices, and presently the parents began to bring back their boys and girls. Even Red Cloud risked his daughter again; but a short time after the reopening of the school he went to the big building on an unannounced visit of inspection and found his little princess scrubbing the kitchen floor, a buxom white cook standing over her threateningly. The enraged chief almost frightened the female staff to death before he went storming off to his camp, bearing his rescued child with him. He at once raised the banner of revolt again, going from one camp to another to make speeches against the agent and his boarding school.

In the big school building the New England women crept about their tasks, their apprehensive looks turning constantly toward the row of windows that looked out straight into the center of Red Cloud's camp beyond the creek. Rumor had it that this camp was crammed with armed warriors awaiting the signal, and every time an Indian yelled the white women in the school building had hysterics. Even McGillycuddy was alarmed at last; so much alarmed that he sent mounted men flying to the Kiyuksa camp forty miles away on Yellow Medicine Root Creek, to summon Little Wound and all his warriors to the defense of the school and agency. As usual in this seven years' war, not a shot was fired.

Mrs. McGillycuddy asserts that this school war was ended in ten days by her husband's intrepid action. The records in the National Archives show that the trouble went on for weeks and that echoes of the strife reverberated for months. Some of the school staff fled in dismay, the school principal, Miss Emma Sickles, taking refuge across the Nebraska line, where some months later she was running a roadhouse, in the older and more respectable meaning of that term.

While McGillycuddy and Red Cloud fought it out on the spot, the newspapers both East and West took up the quarrel, some siding with the agent, others with the old chief. The *Yankton Herald* shouted that McGillycuddy was unfit to be an agent: a cantankerous, quarrelsome, arbitrary, pompous, self-willed fellow. With this view the editor of a Deadwood paper of near date agreed, adding bitterly, "and, besides, he wears corsets." Other Western papers defended the agent with vigor. The editor of the *Northwest Nebraska News* was so stirred up that he became entangled in his own verbiage, expressing the view that the trouble at Pine Ridge was due to the activities of "worthless Indian obstructionists who are too antiquated by at least fifty years." [17] Most of the editors who supported the agent stressed the admirable order and discipline he maintained at Pine Ridge, in reply to which the opposition editors emitted catcalls. They pointed out bitterly that the heavens had been rent twice a year ever since this agent came to Pine Ridge by a mingling of war whoops, and shouts from McGillycuddy with threats of bringing in the cavalry.

This school war of 1884 produced the finest crop of charges and investigations as yet brought on by Red Cloud's squabbles with his agent. McGillycuddy went after the old chief with a sharp stick, petitioning the Indian Office for permission to cut off from the ration rolls all families in Red Cloud's camp that refused to put their children in school. There were seventy-five children in the camp, just over the creek and in sight of the agency school, but the officials were not yet conditioned to the idea of starving Indian families in the name of education, and it was not until October 16, 1885, that McGillycuddy at last obtained the special permission he was demanding.

Meantime there were two investigations at the agency and a third hearing in Washington. A House committee of three members came to Pine Ridge, bringing the sergeant-at-arms of the House with them, whether for protection from Red Cloud's warriors or McGillycuddy's police is not stated. The chairman of this committee, W. S. Holman of Indiana, was inclined to favor Red Cloud; indeed, he picked up some of Red Cloud's friends who had been banished by the agent and brought them with his committee from Nebraska to the agency. McGillycuddy, not in the least awed by a House committee and a sergeant-at-arms, ordered his police to expel these squawmen from the reservation at once, and he then informed the astounded and enraged Holman that anyone who disregarded his agency rules, up to and including the chairman of a House committee, would be instantly removed from the reservation by his police.[18]

The struggle at Pine Ridge might have gone on forever, Red Cloud not strong enough to rid himself of the agent, McGillycuddy lacking sufficient power to break the old chief; but a

[17] *Northwest Nebraska News*, May 14, 1885.
[18] *McGillycuddy Agent*, 240.

national event of first importance now intervened. In November, 1884, the Democrats elected Grover Cleveland as president and swept into power. Red Cloud lifted his head. McGillycuddy was a Republican, and now the old chief and his followers thought the hour had struck.

But the outgoing Arthur administration in a last-minute act reappointed McGillycuddy as agent for another full term of years. This greatly annoyed both Red Cloud and the Democrats, for during the campaign Cleveland had pledged his support to the newest political fad: civil service reform. No Indian agent would be removed for political reasons. But the Democrats had not been in power since 1860 and the rank and file of the party were starving to death for patronage. This was particularly true of the southern Democrats, who had not had a man in a federal position since the Civil War.

Cleveland turned over the Interior Department and its Indian Bureau to southern Democrats, some of whom at once set to work to discover a method for getting around or crawling under the civil service rules and putting an army of Democratic office-seekers into positions held by Republicans. Red Cloud seemed a real answer to prayer to these men, and when the chief began to shout anew against his agent, he was encouraged. He and McGillycuddy were brought to Washington again, and before the new Democratic secretary of the interior Red Cloud denounced his agent. The new officials listened attentively and then gave up hope. Red Cloud's angry talk was that of a child. The same old accusations–McGillycuddy was a thief and tyrant, he starved his Indians, and so on. McGillycuddy stood up and refuted everything the chief said. Dr. Bland and Judge Willard (father of Frances Willard, the temperance crusader) were there, backing Red Cloud; but the hearing was a farce. The chief simply had no case.

The southern Democrats in control at the Indian Office now took matters into their own hands and set about getting rid of McGillycuddy. They sent snoopers to Pine Ridge to search diligently for evidence of wrongdoing, and as the work progressed the Democratic officials began to have a higher view of Red Cloud's mental capacities. No wonder the chief's case against McGillycuddy was so pitifully weak! This agent was armed at every point, with not a chink in his armor through which a weapon might be slipped. His bookkeeping was perfection; his probity above criticism; and not a thing could be found that might be employed as a hook on which to hang charges. The Democrats gave up and resorted to Red Cloud tactics. Make McGillycuddy angry and let him hang himself.

A telegram was sent from the Indian Office announcing that a new chief clerk would soon arrive at Pine Ridge. McGillycuddy reacted instantly and with his customary vehemence. He sent a telegram refusing point-blank to receive the new chief clerk. He stated that hundreds of thousands of dollars worth of public property were in his hands; he was a bonded officer; he was often called away on official business; and it was imperative that he should have a chief clerk of his own choice whom he could trust absolutely. He must therefore decline to accept a new chief clerk, selected for political reasons. So there it was. Disobedience and insubordination, two charges that McGillycuddy had so often hurled at Red Cloud. This was intolerable; it was a defiance of all authority. The Indian Office officials decided to act.

An inspector was sent to Pine Ridge, where he asked McGillycuddy before witnesses if he would accept the new chief clerk appointed by his superior officers in Washington. He replied curtly that he would not. The inspector then relieved him of duty and put Captain James M. Bell of the Seventh Cavalry in as acting agent. Thus McGillycuddy was removed, and he and his supporters were deprived of the opportunity to shout that he was the victim of political jobbery. Captain Bell was a close friend of McGillycuddy; moreover, he was a soldier, and no one could say that by removing McGillycuddy the Democrats had let down discipline at Pine Ridge and risked the chance of an Indian uprising. Captain Bell was kept as acting agent until the short memory of the public forgot McGillycuddy; then a worthy Democrat—an Irish colonel with a good Civil War record—was quietly slipped in as agent at Pine Ridge, and peace descended on the field of battle where Red Cloud and McGillycuddy had fought for seven years.

McGillycuddy remained in the West, being appointed head of the Dakota asylum for the insane. Red Cloud said that he was sorry for the poor mad people. They would have no rest; notice boards would bristle for twenty miles around, and the attendants would be armed and drilled like soldiers.

4
The Adventures of the Three Musketeers

WHILE LEADING humanitarian groups in the Eastern states were exerting pressure in Washington in 1878 for a scheme aimed at making all the Sioux into farmers in three or four years' time, another group of crusaders was at work in Dakota Territory with plans for attracting a flood of white settlers into the Sioux lands. The attitude of these Western men toward the Indians was utterly at variance with that assumed by the Eastern groups of Indian Friends, who regarded the tribesmen as their red-skinned brothers whom they in Christian duty must help. To the Dakota leaders the Sioux were just a horde of lazy barbarians, settled by a foolish government on lands wrongfully taken from the Dakota whites and given to these Indians, who had no intention of changing their ways but would forever remain a hindrance to the progress of white settlement. The purpose of the Dakota men was to push the Indians into a corner, take the best of their lands, and settle white families on them.

Most of the Dakota leaders were fifty-niners who had come into the wild Sioux land with high hopes immediately after the Yankton Sioux had sold all of Dakota lying east of the Missouri to the government by treaty in 1858. They were partly Westerners, but to a large extent they came from good families in the Eastern states. Newton Edmunds, one of their leaders, was an Easterner who came to Dakota as an official of the territorial government and later became governor of Dakota. He and his friends formed an inner circle of Eastern and New England men who constituted themselves the ruling class in the new territory. They attracted a flow of farming families into the vast tracts of empty prairie, directed the flow of immigration, and by advertising Dakota in European countries they drew the first colonies of Bohemians and Scandinavians to the Sioux land. Many of these Dakota leaders were men of superior types. They were of the same blood and upbringing as the leaders of the Eastern Indian Friend groups of the 1870's. But life on the frontier had seriously affected any predisposition of the Dakota men toward idealistic views concerning Indians, and many of them had picked up the border belief that the Indian had been created for white men to fleece and that the United States government was a milch cow provided by a kindly providence to feed a hard-pressed frontier population. Without in the least abating their self-respect some of the Dakota men of the ruling class were adding to their family income by misappropriating government funds provided for the benefit of the Sioux. Their friends and neighbors—as highminded citizens as could be found in any of the Eastern states—were not critical of their rather peculiar methods of making a living. On the frontier, government money was manna from heaven, and no one thought the worse of a man for collecting a few buckets.

These Dakota men had started a land boom almost as soon as they had established a legal residence in the territory, but they had bad luck—the outbreak of the Civil War in 1861 cut down the flow of farming families to Dakota, and the Minnesota Sioux uprising of 1862 with the ensuing Indian war still further retarded the work of settlement. Then came the peace treaties with the Indians. The treaty of 1868 outraged the Dakotans by giving the Sioux all the lands lying west of the Missouri, north to the Cannonball, and this was confirmed by the treaty or agreement of 1876, which took the Black Hills from the Indians but left them with all the rest of Dakota, west of the Missouri. By 1877 most of the good land east of the river had been taken up by settlers, but the flow of farming families was continuing, large groups still coming from Bohemia, Germany, and the Scandinavian countries, with the result that the Dakota leaders and particularly the land boomers were casting longing eyes on the vast tracts of Sioux land beyond the Missouri. They considered it an intolerable situation that the settlement of Dakota should be stopped dead at the river brink, because silly Indian Friend groups in the East and a misguided government were bent on keeping all the land west of the river for the shiftless Sioux, who did not intend to farm, no matter what pressure was brought on them.

The railroads had crossed Dakota west to the Missouri, and there they had stopped. They would not build their lines on westward through Indian country where no one worked and there was no local freight of corn, wheat, and cattle to bring the road at least enough revenue to pay the cost of taking tracks and trains through the district. Here again, as the Dakotans viewed it, the Sioux were standing in the way of progress and killing the growth of the territory.

With no faith in the program of the Eastern Indian welfare groups or in the government Indian policy, dictated in the main by these Eastern groups, the Dakota men looked on, angry and often jeering, while Spotted Tail put on his surprising performances at Rosebud and Red Cloud helped his agent to stage the uproarious battles at Pine Ridge. Eastern people might be thrilled or amused by these acts, but the Sioux were sitting down on 43,125 square miles of Dakota land, and the Dakotans wanted that land for settlement. To them the government Indian farm plans were a bad joke. These Sioux were sitting there, eating free rations, wearing free clothing, with no intention of doing anything else.

As long as Rutherford B. Hayes was president, there was no chance for Western groups to be given a free hand in dealing with Indians, and the new president, James A. Garfield, was a man of the same high character as President Hayes. But early in the winter of 1881 President Garfield was assassinated by a crank and Chester A. Arthur moved into the White House. Arthur was a product of New York machine politics and was reputed to be an easy man to approach. His secretary of the interior was Henry M. Teller, a Westerner whose views concerning Indians were similar to those held by the Dakota leaders. Teller might give lip service to the Indian policy forced on the government by the

Christian and humanitarian pressure groups, but in private he was ready to listen to reason and to give the Dakota leaders what assistance he could.

Almost at once after Arthur became president there were indications of a liberalizing of the government views concerning the opening of more Sioux land to settlement. Before anything could be set in motion, however, Congress had to be brought around, and many members of Congress were too much impressed by the powerful Indian welfare pressure groups to take any open action that those groups might object to. The railroads knew that something was brewing and acted first. Permitted to deal directly with the Sioux, they took advantage of them, purchasing a right-of-way across the Crow Creek reservation for $375 and obtaining a fine large base on the Missouri bank at Fort Pierre for $3,200. The Sioux then woke up and began to shout that they had been robbed.

The next move was made at the opening of Congress in 1882, when R. F. Pettigrew, delegate from Dakota Territory, introduced a bill (H.R. 4630) providing for a commission to go among the Sioux and find out if the Indians were willing to sell their surplus lands. The sponsors of this measure stated later that it had been approved of by all the Sioux agents. In truth the only agent who favored the bill was McLaughlin of Standing Rock, and he cautiously refrained from making a statement until after Congress had enacted the measure. This bill attracted no notice and did not seem to have a chance of passing, but Pettigrew knew his way about, and in some manner he got his little bill pinned to the important sundry civil appropriation bill as a rider, both bills being enacted during the rush and confusion of the closing session.

Even after its passage the Sioux Land Commission Act attracted no attention. H. Price, the commissioner of Indian affairs, had his notice drawn to the act, and he wrote to Secretary Teller to suggest that if a commission was to go among the Sioux, a letter of instructions should be drawn up. Teller agreed, and Price had the instructions prepared at the Indian Office. The selection of the commission was evidently in Teller's own hands. As chairman he picked Newton Edmunds of Yankton, Dakota, a former governor

Hollow-Horn-Bear (Mato He Oklogeca). Reprinted, by permission, from Hamilton and Hamilton, The Sioux of the Rosebud, *pl. 216.*

Courtesy South Dakota Historical Society

Swift Bear and Charles Jordan

of the territory who had been in charge of Sioux affairs for many years and was highly skilled in talking Indian chiefs into signing treaties and agreements. He had persuaded the Sioux to set their marks on sheets of paper many times, and if the Indians had ever benefited in any manner from his attentions, no one knew when or where that had been. Judge Peter C. Shannon, the second member of the commission, had served as chief justice of the Dakota Supreme Court until 1881. In 1877 he had presided over the court before which Jack McCall was tried and sentenced to hang for the shooting of Wild Bill Hickok. Judge Shannon was a typical frontier jurist. He shared to the full the border belief that Indians were worthless creatures, a hindrance to white progress, and that their so-called treaty rights were a lot of nonsense. The third member was James H. Teller of Ohio, who owed his appointment solely to the fact that the Secretary of the Interior was his brother.

Commissioner of Indian Affairs Price had prepared an honest and straight-forward letter of instruction for the guidance of this commission. He informed the gentlemen that the act required that they should go to the Sioux reservation, consult with the Indians, find out whether they cared to sell part of their lands, and if so what lands and at what terms. Instead of doing this, the commission went to Newton Edmunds' office in Yankton and wrote an agreement under which the Sioux gave up almost half their land and just those particular tracts that the Dakota land boomers most wanted. It was a cleverly prepared document, so worded that the simple Indian chiefs would not realize how much land or what places they were giving up. Such small points were hidden behind the proposal to give the Sioux at each agency a splendid new reservation consisting of lands they already possessed, the kindly commission making a gift of these lands to the Indians who already owned them. In effect, up to this time the great Sioux reservation had not been divided; all the Sioux had equal rights to all the lands; but now seperate reservations were to be created for the Sioux at Rosebud, Pine Ridge, Lower Brûlé, Cheyenne River, and Standing Rock, and the gentlemen of the commission evidently hoped that in their pleasure at seeing some boundary lines marked on a map the Sioux would not notice particularly that the whites were taking half their lands. The line of approach the commission intended to employ in speaking to the chiefs was that with the great reservation undivided and held in common, the Sioux had a poor title to their lands, but that by accepting a separate reserve for each Sioux group the Indian title would be greatly strengthened. This was a good talking point, but whether there was any truth in the talk about the Indian title is a very doubtful matter. The Sioux held their lands by treaty, with the honor of the United States government pledging the soundness of that title.

This land agreement exhibited to the full the Dakota men's understanding of the Sioux and their skill in dealing with them. It was not only a very clever document, it was sly. It did not offer the Sioux money for the lands they would surrender; if it had done so, some unscrupulous persons might have told the Sioux that the lands being taken were worth at the very lowest valuation around $6,750,000. In place of money the document offered the Sioux one thousand bulls and twenty-five thousand cows. That was a payment the Indians could understand and that would impress them as being very desirable; but the money-worth of the animals proffered was less than $1,000,000.

While the commission was in Yankton at work on the agreement, Newton Edmunds bombarded the officials in Washington with letters and telegrams. He was adept in Indian treaty negotiations and had all the tricks and small refinements of that art at his finger ends. Congress had appropriated $5,000 for the expenses of the commission; surely enough to pay the way of three men while visiting six adjacent Indian agencies to ask the Sioux if they cared to part with some of their lands, but by no means enough to take care of the very different program that Governor Edmunds was mapping out. Edmunds now wrote to the Commissioner of Indian Affairs asking for money from the Indian contingency fund—a fund of whose existence Commissioner Price (a new official) probably was ignorant. What did Governor Edmunds want the money for? Well, for the providing of feasts for the chiefs and headmen at each agency. This was disapproved by Price, and the request for funds was rejected. Edmunds next asked for the services of a clerk from the Indian Office, to save the commission the expense of hiring one of their own. Disapproved. He asked that the Secretary of War be requested to instruct commanding officers at military posts to place transportation, cavalry escorts, and other facilities at the disposal of the commission. This was approved and saved the commission much money, at the expense of the War Department.

It was apparent from the beginning that Newton Edmunds had no intention of carrying out Commissioner Price's instructions to abide by the provision of the treaty of 1868 and obtain the approval of three-fourths of the adult Sioux males to the new agreement. He was planning what General W. S. Harney had once contemptuously termed "a cracker and molasses treaty" with the Sioux, in which the consent of only a few chiefs and headmen would be sought. That was why Edmunds desired a fund for giving feasts to the chiefs and headmen. The commission's instructions provided that they were to communicate with the Commissioner of Indian Affairs alone; but Edmunds seems to have felt that Commissioner Price would be too stiff a man to handle in this very important matter, and he made his request by telegram direct to the Secretary of the Interior. Secretary Teller, who seems to have considered the matter a small detail, wired his consent to setting aside the three-fourths provision of the treaty. He had no more authority for doing this than he had for altering the terms of our treaty with Great Britain.[1]

Another matter that Governor Edmunds regarded as extremely important was that the commission should have its own Sioux interpreter, loyal to the commission alone. The agency interpreters—all mixed-blood Sioux—would not do. They all had some loyalty to their own people. Edmunds was determined to have as interpreter the Reverend Samuel D. Hinman, who had been a missionary among the Sioux since his youth, who spoke better Sioux than most Sioux did, and who understood these Indians better perhaps than any other white man. Edmunds and Hinman were old companions who had campaigned together among the Sioux and other Indians on many occasions. Hinman had been

[1] *Dawes Report*, 327.

interpreter for the great treaty commission of 1868; he had been a commissioner himself in 1876—one of the praying men who mourned over the fallen tribe and neatly abstracted from it the Black Hills, the Powder River and Bighorn lands. Hinman had been badly singed in the Ponca affair, for he was said to have been a leader in frightening the little tribe into giving up its lands in northern Nebraska. Newton Edmunds had been reported at the time as a leading spirit in that unsavory business.

Hinman had left Dakota and was now employed by the U. S. Census Bureau as an Indian specialist. On October 3, 1882, James H. Teller wired from Yankton to ask his brother, the Secretary of the Interior, for permission to employ Hinman. At the same moment Hinman (on his way to Montana on census business and apparently unaware of what the men in Yankton were doing) wired to Secretary Teller that if his services were required by the Sioux commission, he could be reached at the Merchants' Hotel in St. Paul. Not knowing where Hinman was, the worried Sioux commission rained telegrams into Washington, beseeching Secretary Teller's aid in getting hold of their indispensable interpreter. At the height of the excitement Hinman walked into Newton Edmunds' office (October 14) with his appointment as interpreter in his pocket, approved by Secretary Teller in person. But that did not end it. Hinman knew his value to any commission seeking to induce Sioux chiefs to sign away their lands. Hinman was being offered one hundred dollars (equivalent to about five hundred at the present day) a month and all expenses paid; but he demanded ten dollars a day, the same remuneration that Chairman Edmunds was to receive. The three commissioners almost wept. They showered Secretary Teller with telegrams, carefully avoiding waking up Commissioner Price at the Indian Office (who was certain to veto the proposal), and in the end Secretary Teller gave his assent.[2]

On October 16 the commission left Yankton for Fort Niobrara,

[2] *Dawes Report*, 324–32. Criticisms, often very severe, of Edmunds and Hinman were printed in the Niobrara, Nebraska, *Pioneer* (1878) and in some of the Yankton papers from 1877 to 1884. The Bishop Hare Papers contain similar criticisms of Hinman, as does also the manuscript journal of the Reverend Mr. Cook.

in northeastern Nebraska, to obtain the free use of a comfortable army ambulance in which to travel. Seeking an easy success with which to inaugurate their labors, they now went to the little Santee Sioux reservation on the Missouri in northern Nebraska. The Santees had no direct interest in the lands of the Sioux in Dakota, but by the treaty of 1868 their consent was necessary to legalize the sale of any portion of the reservation. This little Sioux tribe had suffered terrible losses during the outbreak of 1862 and in the following years, and the survivors had all become Christians and farmers. They had little in common with the Sioux of Dakota, taking the white man's view of any matter that came up for decision, and accepting the advice of their church leaders on all important occasions. Their chiefs now signed the paper that the Reverend Mr. Hinman proferred and the commission went its way, with one victory to its credit.

The next step was a long one—a journey of over two hundred miles across empty plains and hills to Pine Ridge. The party went by way of Rosebud Agency, where they were warned that McGillycuddy of Pine Ridge was engaged in a violent struggle with Red Cloud, that most of the Indians sided with that chief despite McGillycuddy's efforts to put Young-Man-Afraid-of-His-Horse in as head chief in Red Cloud's place, and that if the land commission accepted McGillycuddy's hospitality the Indians would regard them as his friends and would have nothing to do with the land agreement. Meantime McGillycuddy had sent a neat conveyance and a guard of honor from his Indian police, to bring the commission from Rosebud to Pine Ridge. The commission disregarded the warning, accepted the carriage and escort, and on reaching Pine Ridge became the agent's house guests.

On Monday, October 23, three to four hundred Oglalas—chiefs, headmen, and warriors—crowded into the council hall at Pine Ridge to hear the message that these three white men had brought from the Great Father. The commission had written instructions to ask the Sioux if they wished to sell land, how much land, and on what terms. They also had their own handiwork: an agreement prepared in minute detail, providing that the Sioux give up some eleven million acres of the commission's own selection at an absurd price which the commission itself set. In opening this first great council they ignored their instructions; they also apparently concealed the true nature of the agreement, speaking mainly of Pine Ridge affairs, explaining that the Oglalas of this agency had no fixed boundaries and a rather doubtful title to their lands (and here they seem to have glanced at the possibility of the tribe being removed to Indian Territory); whereas, if the agreement was signed, the Oglalas would have a fine big reservation, an excellent title, fixed boundaries, and a large additional strip of land on the east which up to now had been considered a part of the Rosebud reservation. This was clever talk. If any mention was made of the huge tracts of lands the Indians were to give up, that point was passed over lightly and swiftly, but much was made of the fact that the Oglalas were to have thousands of free cows and over one hundred fine bulls. The Indians should have been very pleased, especially as they were getting in on the rival tribe at Rosebud by annexing part of that tribe's land; yet when the commission had ended its talk, the chiefs would not commit themselves further than to state that they would council with their people and give their reply to the commission at a later meeting.

The consultation the Oglala chiefs held with their people lasted from October 23 to October 26. Meantime the land commission who, unlike the Sioux, did not dwell in a piece of Eternity where time had no significance, grew more and more restless and worried. Their Man Friday, Hinman, was working like a Trojan amongst the quarreling Sioux; Agent McGillycuddy, with orders from Washington to aid the commission in every way, was also sending his men amongst the groups of arguing Indians, and he had himself drawn a big map of the Pine Ridge reservation as provided in the agreement which he explained loudly to groups of Sioux, who watched beady eyed whilst their father the agent rapped with a long stick those points on the map to which he was calling attention. Over and over again he boxed the compass, naming all the streams around the borders of the new reservation, jabbing at the map with his stick. He, like the land commissioners, seems to have confined his talk mainly to a description of the new reservation the Pine Ridge Sioux were to have, glossing over the fact that the agreement gave the whites nearly half of the Sioux lands.

When the Indians reopened the council at the agency on the morning of October 27, they chose Red Cloud to speak the tribal will, ignoring McGillycuddy's candidate for head chief, Young-Man-Afraid-of-His-Horse. The old chief stood up and, speaking gravely, informed the commission that his people would not sign the agreement, that they had no land for sale, and that the Sioux at the moment were considering the formation of a federation of all their tribes, one principal object of this union being the prevention of the sale of any more land. Red Cloud, having spoken to the white men, turned to his own people and in an earnest speech advised them to stand fast by their decision not to sign the agreement.

The agreement seemed lost; but Newton Edmunds knew his way about, and he now turned to Agent McGillycuddy and asked him to speak to his Indians. McGillycuddy (as he reported to the Indian Office on November 11, and with evident pride in his action) instantly sprang to his feet and addressed the Sioux "in a somewhat forcible manner, instructing and advising them" to hold another council among themselves next morning. He added that the commission would receive their answer the following afternoon and would expect to hear from all of them—chiefs and commoners, young men and old.[3] These Indians had already given their answer to the commission; but McGillycuddy was pleased to pretend that Red Cloud, speaking the will of the tribal council, did not represent the Indians. McGillycuddy was posing again as the father and friend of the tribe: a role familiar to the Oglalas, who detested it. But they feared this man in whose hands rested the power of the government; yet, after endless counciling and quarreling, with the agent exerting pressure to obtain his will, the Sioux again decided to reject the land agreement.

Bishop William H. Hare of the Episcopal church was present at Pine Ridge and was appealed to by Newton Edmunds to employ his influence to swing the Indians into line. He declined to intervene. Someone was circulating ugly reports, seemingly the

[3] *Dawes Report*, 334.

same old threats that if the Sioux did not give in, troops would be sent against them and they might all be packed off to Indian Territory. Slowly the Indians, a few at a time, were frightened into deserting Red Cloud and the tribal council, and when the meeting in the agency council hall reassembled Saturday afternoon, it was reported that a large number of the Sioux had been brought around and intended to let their names be signed to the agreement. The session had hardly gotten under way when Red Cloud rushed forward and seized the pen, as if it were the cross of salvation, insisting that his name should be signed first. Other chiefs and headmen followed swiftly, and seventy-nine names in all were put down. Newton Edmunds then announced that the Oglalas of Pine Ridge (population 7,200) had accepted the agreement. What his commission had told the Indians that the agreement meant is made clear by the message he now handed in at the agency telegraph office. It was to the Indian Office, and it said: "The chiefs and headmen of this agency have this day unanimously agreed *to a separate reservation* with good feeling and satisfaction, Red Cloud and his friends joining in." Here there is no hint that 17,000 to 18,000 square miles of Sioux land had been signed away.[4]

Newton Edmunds had good reason for being pleased by the results at Pine Ridge. Here was proof that the Sioux had not changed much since the days in 1876 when he and Hinman and their fellow commissioners had prayed over the Oglalas and told them that they were in terrible danger, but their good friends of the commission would save them. Even Red Cloud had believed them and had let them sign his name to the paper which took from the Sioux the Black Hills and the lands in the Powder River and Bighorn country. The Sioux were brave enough; but ever since they had come to the reservation they had been like wild creatures in a cage, ill at ease, watchful, and subject to sudden fits of panic. Men like Edmunds and Hinman knew how to play on their fears by sly talk about sending the Sioux to Indian Territory where most of them would die, about cutting off rations and letting them starve. They talked until the Sioux were dizzy, confused, and frightened, then led them like sheep and affixed their

[4] *Ibid.*, 332.

names to the new agreement. This was an approved frontier method for dealing with Indians, and it was about as ethical as confusing and frightening small children and then robbing them of their little treasures.

The treaty of 1868 said distinctly that no future agreement involving Sioux lands should be effective unless it had been approved by three-fourths of the adult males. Commissioner Price had instructed the Edmunds commission that it must abide by this rule; yet Edmunds reported the triumphant approval of the agreement after only 79 of the reported 7,200 Oglalas had been talked or frightened into signing. The three commissioners had obviously decided to ignore the treaty obligations, and after the Pine Ridge victory they felt certain that it would be easy to obtain at each agency enough signers to warrant a claim that the land agreement had been approved by the Sioux. The gentlemen were so confident that they sent some communications to Yankton, and the newspapers of that town blossomed out with glowing descriptions of the immense tracts of Sioux lands that would soon be open to white settlement. These newspapers reached Pine Ridge while McGillycuddy was still glowing over his victory in forcing his Sioux to change their minds and approve the agreement. The squawmen read the papers and communicated their contents to the chiefs and headmen, and then the camps began to buzz like hives of angry bees. Red Cloud and most of the other signers at once repudiated the agreement. No one had explained clearly about all this land being taken; it was another swindle. Councils were held and messengers were sent off to the other agencies, to urge the Sioux to refuse to sign the paper.

Meanwhile the land commission had made its way eastward to Rosebud. Here there was a weak agent, John Cook, from whom the commissioners could expect little real help. Moreover, the Brûlés had held a council before the commission arrived, had voted to reject the land agreement, and had chosen White Thunder of the Loafer Band to speak the will of the tribe. This news only made the commission smile. They set to work cheerfully on the Brûlés—united in opposition. The commission gave feasts to put the Sioux in a good humor; they hired squawmen and mixed bloods to talk to the Indians until they were too dizzy to think and would do as they were told. Some of the Brûlés volunteered to help. Swift Bear, chief of the Corn Band (now old and turned Christian and an admirer of the whites) did all he could to obtain signers. Young Spotted Tail was courting the land commissioners, perhaps thinking that they would aid him in achieving his ambition to be head chief. He was living in his father's mansion, where he was very uncomfortable; but that, he hoped, would add to his prestige in the tribe and impress the whites. He now wore citizen's clothing and rode about in a fine buggy. Yet the tribe put up his rival, White Thunder, to speak for it.

Brown Hat (alias Baptiste Good) was also trying to exhibit his importance. This capacious windbag, so often pricked and suddenly deflated by old Spotted Tail, was now trying to set himself up as a chief. He claimed to be educated, telling a tale of his adventure when a youth; he claimed that he had gone underground into a huge beaver house and had there met four mysterious strangers, who taught him to read and write in four days. Many of the Brûlés believed it. After all, was he not writing a history of their tribe in the form of colored pictures, so wonderful that Sioux who could not read a word in a white man's book could read and understand Brown Hat's picture record?

When the land commission had done what was possible to soften up the Brûlés, a council was called and the agreement was explained by the interpreter, Hinman. Here, as at Pine Ridge, emphasis was placed on the fact that the local Sioux would have a fine reservation if they approved the agreement, and little was said about the giving up of half the Sioux lands to the whites. A year later Herbert Welsh of the Indian Rights Association visited this agency, seeking the facts, and White Thunder assured him that the Indians were not told by the commission that any large quantity of Sioux land was involved in the agreement. The Reverend W. J. Cleveland, missionary at Rosebud, gave Welsh the same information. And when the Dawes committee questioned Agent Wright and his chief clerk, these men who had been present at the Edmunds councils stated that they heard nothing said about the sale of any large tract of land.[5]

In the final council, White Thunder, speaking for the tribe, stated that the Indians rejected any proposal for a land sale. He was ignored; the land agreement was laid on the table, and the commission's paid agents led the chiefs up to touch the pen and watch their names being written down. The chiefs protested that they did not wish to sell any land, and yet they took no firm stand against the signing. It seems certain that none of them knew the true nature of the agreement and that they were probably influenced by the fact that the Oglalas of Pine Ridge had already signed. One hundred and nine chiefs, headmen, and common Indians signed, including twelve squawmen and mixed bloods. The commission then announced that the Brûlés of Rosebud had approved the agreement.

Having obtained what they were pleased to term the willing consent of the Sioux at the two great agencies, Pine Ridge and Rosebud, the land commission decided that their task was more than half accomplished and that they might now return to civilization at Yankton for a little needed rest. This was a tactical error, for in Yankton someone talked too much, and the newspapers of the frontier began to sing about the Sioux lands the commission was to obtain for white settlement. The squawmen on the reservation read the papers and spread the information thus obtained, and most of the Sioux became very much alarmed. Red Cloud was preaching a holy war against Edmunds and his land agreement, and every day the Sioux opposition to the agreement grew stronger.

This storm broke in November; but before that the commission left Yankton by steamboat for the most northerly agency, Standing Rock. The agent here, James McLaughlin, was the only Sioux agent who had approved the land agreement in an official report, but even he had warned that no attempt should be made to get around the provision of the treaty that three-fourths of the adult males among the Sioux must approve of any sale of tribal lands. He was now called upon by his Washington superiors to do all in his power to aid the commissioners, who were openly ignoring

[5] *Ibid.*, 165. Herbert Welsh, *A Visit to the Sioux Reservation* (Philadelphia, 1883), 26–27.

the treaty stipulations. McLaughlin followed his orders, although he did not approve of the commission's methods of seeking signers, and he considered the compensation the Indians were to receive for their lands both inadequate and improper. Newton Edmunds and his two fellow commissioners had put it into the agreement that the one thousand bulls and twenty-five thousand cows the Sioux were to receive were to be American cattle taken from midwestern farms, where the animals had been sheltered in barns and fed in winter. McLaughlin pointed out that in the Sioux country these animals would be turned loose to shift for themselves on the open range, and one Dakota winter would destroy most of the compensation the Indians were to have in payment for their lands. Unfortunately, his sense of duty prevented his speaking his mind in open council while the Edmunds commission was dealing with his Indians.

It was known to everyone concerned from the start that the Standing Rock Sioux were united in opposition to the sale of any land. These Sioux had taken a pledge from their chiefs and headmen not to sign the agreement. The Standing Rock Sioux had been in contact with the whites longer than the Rosebud and Pine Ridge Sioux; they were really attempting to farm, and they had a deeper feeling against the sale of any more land than the wilder Sioux had. They also felt more strongly the government's failure to carry out some of the terms of the treaty of 1868. Under that treaty every Sioux man who tried to farm was to receive a cow, a yoke of work oxen, and certain farming implements. The government had done nothing; and the Standing Rock Indians were saying bitterly that if the treaty had been honestly executed, they would have cows and oxen and would have no reason for selling more land to the Edmunds commission for cattle. The Standing Rock Sioux were the only people in the tribe who wanted schools. While the Pine Ridge and Rosebud Indians were fighting against the establishment of schools, at Standing Rock, the chiefs complained that the treaty promised a day school for each thirty children, and they said that under the treaty their group was entitled to twenty-five day schools. They had none. The boarding schools at Standing Rock were regarded as Roman

Catholic church property. The government had built them and had hired staffs of Benedictine brothers, lay brothers, and nuns; but somehow the Indians and even the whites thought the Roman Catholic church had built and was running the schools.[6]

McLaughlin was liked and trusted by his Sioux; but now when he began to coax and persuade them to sign the obnoxious agreement, they were greatly perturbed. To render their position even more unhappy, White Hat (Bishop Marty, head of the Roman Catholic diocese of Dakota) was at the agency and, instead of refusing as Bishop Hare of the Protestant Episcopal church had done to interfere in this secular matter, he took a leading part in the campaign to induce the Sioux to sign. At this agency the commission employed the same tactics it had used at Pine Ridge and Rosebud, refusing blandly to see that the Indians were opposed to any land sale, and utilizing every possible form of pressure to break down their opposition. Under the strain of the steady pressure exerted on them, the Sioux became bewildered and frightened. On the last day of the long struggle they gave way suddenly to one of those fits of panic the Sioux of that period were subject to, and rushing to the table, they almost fought for a chance to touch the pen and have their names written down by the Reverend Hinman. And these were the warriors whose whirlwind assault had thrown back Crook's strong column in June, 1876, and in the same week had overwhelmed Custer's command! Three black-coated civilians, one bald, one wearing a wig, and a Roman Catholic bishop, had brought the Sioux to this shame—simply by talking them dizzy!

At this agency Chief Grass was the agent's pride. In twenty years Grass had turned himself from a wild Sioux hunter and warrior into a hard-working Christian farmer. He had good sense and he spoke the truth. His testimony before the Dawes committee in 1883 is most interesting, as showing why the Standing

[6] There were three boarding schools at Standing Rock, a school for boys and one for girls at the agency and a farm school for boys some miles to the south. They were staffed by Austrian Benedictines and Austrian nuns and lay brothers, most of whom could not speak English. An Irish priest headed the boys' school and an Irishman the girls' school.

Rock Indians—all opposed to any sale of land—suddenly rushed up in panic to sign the Edmunds agreement. Here it is:

QUESTION (BY SENATOR DAWES): What did you think that the agreement about giving up land to the Great Father meant?

ANSWER (BY CHIEF GRASS): Those men talked a great deal, and we were bewildered. It was not with willing hearts that we signed the paper.

Q.: Didn't these men tell the Indians that the paper meant that they should give up a part of their reservation for some cows and bulls?

A.: Yes; that is what they told us.

Q.: That if they would give up a part of their reservation they should have some cows and bulls?

A.: We never said that we would give up any of our land and take cows and bulls for it.

Q.: Didn't you mean to agree to it when you put your name to the paper?

A.: I did not.

Q.: What did you mean by putting your name to the paper?

A.: These men made my head dizzy, and my signing it was an accident.

Q.: How did they make you dizzy?

A.: There was so much talk by Indians and white men that I did not stop to consider whether it was right or wrong, but signed the paper without giving it the proper thought.

Q.: What did these white men say to you to get you to sign the paper?

A.: We talked with them about the land they wanted, and they said they would give us cows; and I argued with them because I was looking to the benefit of my people in the future, and I wanted them to keep their land. When I was talking to them the Indians all at once rushed up from behind me and signed the paper, and everything was so mixed up I didn't know what I was doing until after I had signed the paper.

Q.: Did you expect to have the land and the cows both?

A.: No; we wanted nothing but our land.

Q.: You didn't want the cows at all?

A.: Of course we preferred to keep our land. It would not be right to expect anything else.

Q.: Which would you rather have, 25,000 cows or the land?

A.: We prefer the land.

Q.: What made the Indians behind you rush up to sign the paper if they did not want to give up the land for the cows?

A.: The white men talked in a threatening way and the crowd of Indians behind me got frightened and rushed up and signed the paper.

Q.: What was the threatening way in which they talked to the Indians?

A.: The man for whom we have a higher regard than for any other white man is Bishop Marty; and he stood before us and told us that if we did not sign it we might as well take a knife and stab ourselves; the consequences would be equally bad. That is what frightened the Indians; and he told us, also, if we did not sign the paper we would be displeasing God. If you were talking for your people, and they got frightened and committed an act because they were frightened, how could you stop them?

Q.: Do you think that is the reason so many Indians at Standing Rock signed the paper?

A.: All these men here present know that was the reason why they went up and signed the paper.

Q.: What did the commissioners say to frighten the Indians?

A.: They told us they could send us away from here to a different country [Indian Territory] if they wished to do it. Three men talked a great deal, but I cannot remember in detail all they said.[7]

Later, when the Edmunds commission was fighting with its back to the wall, Newton Edmunds asked Bishop Marty if any pressure or threats had been employed at Standing Rock, and the bishop replied in a letter that none had. Perhaps his definition of pressure and of threats differed from that of Chief Grass. One can only say that although the Sioux were subject to sudden panic, it required severe pressure or threats to produce one of these panics, and everyone admitted that the Standing Rock Sioux had stampeded in frantic fear to the table, to seize the pen and have their names set down. At all of the other Sioux agencies pressure and threats had been used by this commission. No one had mentioned Indian Territory for years until the Edmunds commission went to work; then suddenly the report spread over the Sioux

[7] *Dawes Report*, 64.

reservation that if the agreement failed the Sioux would be sent south. Chief Little Wound, second in rank to Red Cloud at Pine Ridge, said in 1883 that he had signed the agreement because he did not wish to see his people removed to Indian Territory.

The Standing Rock Sioux signed the agreement on November 30, with 134 Indian names, the largest number obtained at any of the agencies. Sitting Bull did not sign. Matilda Galpin, fullblood Sioux and the only woman chief in the tribe, signed. She had been made a chief by her own people in 1875, for her heroism in saving the lives of the Black Hills commission which had come out, like Newton Edmunds in 1882, to get the Sioux to sign away their lands and had been met by thousands of infuriated warriors armed with rifles. How times had changed! In 1882 three elderly white men, aided by a Catholic bishop, had by words alone frightened the Sioux nearly to death.

When the commission reached Cheyenne River Agency, south of Standing Rock, they found a hostile atmosphere thick enough to cut with a knife. By this time the Yankton newspapers had been received and read at all the agencies, and the Sioux knew the true character of the agreement. Not only Red Cloud but the Protestant missionaries were now circulating the warning that if the Sioux signed this paper, they would be robbed of half their lands. The Cheyenne River Indians did not need much stiffening. They had had an experience with the railroads a year or two before, men coming with a paper to sign and the Sioux waking up later to find that they had deeded away part of their land. Besides, the Cheyenne River people knew from the reports in the newspapers that they had been singled out as the principal victims of this Edmunds land deal. The whole of their present reservation was to be taken by the whites and the Indians were to be removed bodily north of Cheyenne River, into a region where they knew that there was little water and where much of the land was sand and gravel. Buzzing like angry bees, they held councils before the commission arrived and instructed the chiefs and headmen not to touch the pen, no matter what these white men might say or do.

Newton Edmunds and his comrades took their coats off and went to work on this mass of sullen Sioux. They knew how to handle Indians who had made up their minds absolutely not to do what was required of them. The fur flew. According to the Reverend Tom L. Riggs, Congregationalist missionary at Cheyenne River, the chiefs in open council rejected the land agreement, over and over and over, and all their rejections were ignored by the commissioners, who simply came back at them more vigorously than ever from a different direction. The commission had a small army of paid agents (squawmen and mixed bloods in the main), who worked assiduously, employing falsehoods and threats to bring the Sioux around and induce them to sign. These men made great play with the fact that James H. Teller of the commission was the brother of the Secretary of the Interior. They told the Indians that this man's brother stood next to the Great Father; he had the power to destroy Cheyenne River Agency, to set the Sioux adrift on the prairie without food or clothing, and this he would probably do if they did not sign. If they remained stubborn they would lose their land anyway and get nothing for it; for the Sioux at the other agencies had put down enough names to approve the agreement without any Cheyenne River names being added. This was a deliberate lie, but the Cheyenne River Indians had no means of knowing that. Shifting from threats to promises, the commission and its agents tempted the Sioux. Captain W. N. Sage of the regular army, who was present during these stirring days, told the Dawes committee in 1883 that the twenty-five thousand cows and one thousand bulls in the agreement were only a small part of what the commission promised the Indians, if they would sign the paper.

For two interminable weeks of big councils and endless meetings of small groups the Cheyenne River chiefs and headmen stood firm; then the strain began to tell and some of them wavered. In a final council the commission practically imprisoned the chiefs and headmen and kept them under terrific pressure until long after midnight. Some of the Sioux broke down. It was later testified that these Indians were literally dragged to the table by the commission's agents and were held there, protesting, while their names were set down. This was a big agency, yet only thirty-five signed the agreement, the smallest number obtained at any of the agencies.

Exhausted by their weeks of strenuous effort at Standing Rock and Cheyenne River, the commission boarded a steamboat the moment the Cheyenne River Indians had signed the agreement and went down to Yankton for a period of rest. They believed that their labors were now successfully concluded, for they had many reasons for supposing that the Sioux at the two small agencies, Crow Creek and Lower Brûlé, that still remained to be visited would sign without much protest. Under the agreement the Crow Creek reservation was to be cut down in size; but the Sioux there were so alarmed over the whites crowding in on them and threatening to take all of their lands that they would certainly sign the paper, to obtain a good title to what lands were to be spared to them. As for Lower Brûlé, the Indians there were to be removed bodily and all their lands given to the whites, but the commission seemed to be confident in its ability to induce the Lower Brûlés to sign.

This commission seems to have regarded the great Sioux reservation as a closed and soundproof compartment in which they could accomplish their work without the world hearing anything about their methods. In the beginning of their campaign this had been actually the situation; not a word as to their remarkable performances at Pine Ridge and Rosebud had seeped out until they went to Yankton and were indiscreet enough to talk where newspaper men were within hearing. By November the situation was greatly altered. Bishop W. H. Hare of the Episcopal church had written letters to certain influential persons in the East, protesting the methods employed by the commission in obtaining Sioux signatures, and several of the missionaries were also writing letters. All of the leaders of the Indian Friend groups were now awake and watching, and certain members of Congress were beginning to cast ominous glances in the direction of Secretary Teller and his Sioux commission. But it was from a totally unexpected direction that the first bomb was tossed.

In the spring of 1878, on reports that the "Dakota Indian Ring" was engaged in plundering the government and the Sioux at Crow Creek and Lower Brûlé, the Interior Department had requested the War Department to send an officer and a force of troops to

Crow Creek, to seize the agent's office and examine his papers and accounts. Captain William G. Dougherty, First Infantry, was given this duty, which he performed so effectively that Agent Henry Livingston, a Newton Edmunds appointee, was indicted on several counts. J. R. Hanson, another Sioux agent who had aided Edmunds in dealing with Indians since 1865, was also indicted. These agents were acquitted by a Yankton jury, but they lost their agencies, and Captain Dougherty was appointed agent for Crow Creek and Lower Brûlé. He was a good agent; but on September 1, 1881, he was removed to make way for a political appointee, George H. Spencer.

Spencer seems to have owed his position to Secretary Teller, who acted on the advice of Dakota politicians. This agent's coming signaled the beginning of a campaign that was evidently intended to frighten the Crow Creek Indians so much that they would sign the land agreement. White settlers pushed into the borders of Crow Creek and the Dakota papers began to print demands that the reservation should be abolished and the lands given to the settlers. The only friend the frightened Crow Creek Indians seemed to have left was their Episcopal missionary, Reverend Heckaliah Burt, one of God's good common men who had devoted his life to the service of these poor Sioux. It was on his suggestion that the Crow Creek chiefs (April 15, 1882) signed a petition beseeching the Great Father for protection from the Dakota whites, who were every day clamoring more loudly for the Indian lands. The Great Father—Chester A. Arthur—ignored the plea of these frightened Indians, but someone in Washington was annoyed at Agent Spencer (perhaps at his permitting his Sioux to send such a petition) and he was replaced by W. H. Parkhurst. Moreover, Crow Creek agency was closed. Parkhurst was made agent for both reservations, and he established himself at Lower Brûlé, leaving only a clerk at Crow Creek. This was a fairly clear warning to the Crow Creek Sioux that their reservation was in danger of being abolished if they did not do exactly as they were ordered.

On duty with his regiment in distant Arizona, Captain Dougherty continued to take an interest in his Crow Creek and Lower

Brûlé Sioux friends. The Reverend Mr. Burt wrote to him from time to time, and the captain subscribed to some of the Yankton newspapers. From the moment the Edmunds land commission took the field Captain Dougherty watched its operations narrowly, and at about the moment when the commission returned to Yankton after its visits to Pine Ridge and Rosebud, Dougherty in Arizona reached certain conclusions. He smelled fraud and coercion in the commission's conduct; he was as shocked as Bishop Hare, as angry as Red Cloud, and being in a temper and unmindful of the respect due to honorable members of a land commission, he sat down at his desk at Fort Apache and wrote a hurried note of warning to Alex Rencountre, his old mixed-blood interpreter at Lower Brûlé, a man Dougherty had worked with for years and whom he regarded as more intelligent than any of the other Sioux at the agency. Wishing to impress Rencountre with the need for instant action, Dougherty employed rude language that hit straight and hard. He wrote:

MR. RENCOUNTRE: Your people are without one friend, and are going to be robbed of their land, and will not get anything for it, and then they will be disarmed, and may be sent to the Indian Territory. If your people give up their land now there will be nothing left to them, and the Government may do what it pleases with the people, and if they kick, the rations and blankets will be stopped, and starvation will compel you to move.

Your agent has no life, and does not care for you, and is only after what he can take, so you should send word to all the Indians at once to stop the sale of the land and send the commissioners home. The chief man in that commission [Newton Edmunds] is the chief robber of the Indians for the last fifteen years. He lied to the Yanktonais and Two Kettles at old Fort Sully in 1865, and he divided with Livingston and Hanson[8] when they robbed you for over ten years.

Send your people to advise all the other Indians to stop the council about the land, and get an honest man to take your side.

Be quick or you will be too late. Call your wise men and have a council at once.[9]

[8] Henry Livingston and J. R. Hanson, agents at Crow Creek and Lower Brûlé, appointed through Newton Edmunds' influence; indicted by a federal grand jury, acquitted by Yankton juries, but replaced as agents by Captain Dougherty.

[9] *Dawes Report*, 337.

Alex Rencountre was still interpreter at Lower Brûlé. He may have loved the Sioux; but he had his own interests more at heart, and he now handed Captain Dougherty's letter and the newspaper clipping that accompanied it to Agent Parkhurst. Dougherty in his haste and anger had neglected to sign his letter, and this gave Agent Parkhurst the opportunity to refer to it as an anonymous letter written in a disguised hand. Parkhurst seems to have been an honest man who was trying (rather weakly) to help his Sioux; but to his thinking it was little less than sacrilege to openly oppose a land commission which was backed by the great Secretary of the Interior. He was very angry over Captain Dougherty's references to himself, and he sent the captain's letter and clippings to the Indian Office with a letter of his own denouncing Dougherty. In Washington an effort was made to stop Dougherty; but he was off on a one-man crusade to save the Sioux lands and nothing short of a curt order from his military superiors could stop him. He was now writing directly to the chiefs and even sending them telegrams. No Sioux chief since the start of time had ever received a telegram, and they handled these mysterious yellow sheets as cautiously as if they had been explosive. They very nearly were.

Parkhurst had uncovered another villain, finding to his horror that the Reverend Mr. Burt had sent a wire to Dougherty, asking him to come to Crow Creek and Lower Brûlé to lead the fight against the land agreement. Parkhurst, who should have known better, made himself believe that the old missionary was planning to have Dougherty seize the two agencies and perhaps murder him in his bed. Everyone who knew the meek old missionary must have smiled at the suspicion that he could hatch such a scheme.

When Congress met on December 19, 1882, the air was filled with ugly rumors concerning the operations of the Sioux land commission. In an upset election in November the Democrats had won control in the lower house. The new members of the House would not be sworn in until March, 1883; but there were already strong indications that the coming session would be featured by Democratic efforts to blacken the record of the Republican administration. If there was a scandal in Dakota, it behooved the

Republicans to look into matters and try to set things straight, for their own well-being if not for that of the Sioux Indians. The Arthur administration was silent as the grave and made no move one way or the other.

With matters in this position a resolution was passed by the Senate, calling on the Secretary of the Interior for all papers concerning the Sioux land agreement. This was ominous. On January 3, Secretary Teller wired to Newton Edmunds at Yankton advising him to go to Lower Brûlé and Crow Creek at once and wind up the agreement as rapidly as possible. The object of both Edmunds and Teller was to remove the Lower Brûlés *en masse* to the Brûlé reservation at Rosebud and open the whole Lower Brûlé reservation to the whites. It was pretended that this was to be by agreement with the Indians concerned. If, however, there was any intention to consult the wishes of the Sioux, it was very carefully concealed.

The commission left Yankton for Lower Brûlé three days after receiving Secretary Teller's warning, in the evident belief that the Sioux there would sign the agreement at once. The commissioners soon realized their error. Captain Dougherty's letters and telegrams had had a deep effect on the chiefs; the mixed bloods had been reading the newspapers, keeping the chiefs informed on many details of the land agreement that the commission undoubtedly would much have preferred the Indians to be in ignorance of; and, all in all, the feeling among the Lower Brûlés was that they were to be robbed and then removed to lands which were not half as good as the ones they now held. Besides, the news from Rosebud was to the effect that the Brûlés there had no intention of giving any of their lands to the Lower Brûlés, who would therefore be homeless.

The Sioux at Lower Brûlé were conservative folk who had resisted all efforts of the government to destroy their chiefs and put the people completely in the hands of an appointed white agent. Now, in January, 1883, Agent Parkhurst obeyed his orders from Washington and tried to induce his Indians to sign the agreement; but, standing back of their chiefs, the entire tribe refused to have any dealings with the commission. Forgetting the excellent service Parkhurst had performed for them in uncovering the Captain Dougherty and Reverend Burt conspiracy, the commission now reported the unhappy man to Washington as an agent who was so weak that he had no control over his Indians. They then took their coats off and set to work to show the Lower Brûlés who were masters. They swiftly learned that the masters were not present in their own ranks. The Indians stood like a rock, grimly holding out against sweet cajoleries, handsome promises, and angry threats.

Altering their first plan, to just force the Sioux to sign on the line, the flustered gentlemen tried to negotiate. Since the plan to take the Lower Brûlé lands and give these Indians harborage at Rosebud had fallen through, because the Rosebuds would not give any of their land to the refugees, a new proposition was put before the chiefs. The Lower Brûlés would give up their present lands and would receive in return a tract even larger, lying south of White River, extending from the Missouri to the east line of the new Rosebud reservation. The chiefs would not even talk of this plan. The commissioners took time out to consider this shocking impasse. Judge Shannon (one of the commissioners) dipped into his lore of legal stratagems and suggested a clever scheme for adding two tiny amendments to the agreement which would make it possible to take the Lower Brûlé land without the necessity of hammering these plaguey Sioux into signing the document. The situation at Lower Brûlé seeming hopeless, even the upright Newton Edmunds was won over to this bit of chicane, and J. H. Teller was dispatched to Washington, bearing to his brother a document which the commission was pleased to call a completed and approved agreement with the Sioux, although not one Lower Brûlé or Crow Creek Indian had signed it. Newton Edmunds also went away; but Judge Shannon and the commsision's high-priced interpreter, the Reverend S. D. Hinman, remained to see if something could not be done to budge the Lower Brûlé chiefs out of their mulish refusal to sign.

In Washington, Secretary Teller, who was in a great hurry, approved the Sioux agreement with the Shannon amendments attached to it; two days later (February 3) President Arthur also approved the agreement, but removed the amendments, which his legal advisers pronounced improper. That was a mild description of them. The agreement was now submitted to the Senate for final approval. While the Senate busied itself with affairs more important than Sioux land deals, Judge Shannon and Hinman continued to labor with the stubborn Lower Brûlés. These Indians had held out for weeks in the face of every form of pressure and threats; but there were signs now that they were beginning to weaken. They were frightened half to death, and on the verge of panic. Still, led by the head chief, Iron Nation, the Lower Brûlé chiefs held their tribe firm. Yet they knew that they could not do so much longer. Growing desperate, they planned to send two or three of the chiefs to Washington, to make a personal plea to the Great Father for justice and protection. These simple Indians had faith in the man who occupied the White House; but they had no money, and to raise a little fund to pay the expenses of the journey East the Indians planned to sell beef hides to the agency traders for cash. On February 10, Judge Shannon and the Reverend Hinman sent a telegram to J. H. Teller in Washington, asking him to obtain from his brother, the Secretary of the Interior, an order stopping the sale of beef hides at Lower Brûlé and forbidding any chief to leave the agency.[10] Their purpose was to prevent these frightened Indians from going to Washington to appeal for protection. In all our official dealings with the Sioux there is nothing remotely to be compared with this for despicable meanness. If Secretary Teller ever replied to this amazing communication, his answer has not been found; but no Sioux chief left Lower Brûlé.

J. H. Teller soon came hurrying back from Washington with an "ultimatum" from Secretary Teller to the Lower Brûlé chiefs. This queer declaration stated that enough names had been obtained at the other agencies to put the land agreement through without a name from Lower Brûlé, which agency would probably be abolished and the Indians moved elsewhere, by troops if necessary; that if the chiefs signed, their people would be compensated for the improvements on their land; if they did not sign, their people would receive nothing and would be a lost tribe without

[10] *Dawes Report*, 361.

treaty rights who could be moved wherever the government chose to order, and then removed again and again at the will of the officials.[11] A truly enlightening message from the great Secretary of the Interior to a group of frightened Indians, exhibiting to the full how far he was ready to go in forcing this detestable land grab on the Lower Brûlé Sioux. His statements in this message were false, his threats were hollow, for he had no authority to do any of the things he was threatening to do. The Lower Brûlés believed he could; yet when Judge Shannon and the Reverend Hinman quoted these dire threats to the chiefs, the indomitable Sioux refused to sign.

While operating on the Lower Brûlé Sioux, the land commission was also dealing with the Crow Creek Indians, just across the river on the east side of the Missouri. At Crow Creek these men made full use of the Presidential order threat, pretending that President Arthur had authority to abolish the reservation at any time, as it had been set up by a Presidential order and could be destroyed by the same method. This was a notion originated by the Dakota land boomers, and it was ruled against by the United States Attorney General some years later; but in 1882–83 it was a potent threat to hold over the heads of the frightened Crow Creek Sioux. Following the lead of the land commission, the Dakota newspapers started a campaign, predicting that President Arthur was about to order Crow Creek opened to white settlement, and the commission's effort to scare the Sioux into signing the land agreement was greatly aided by a swarm of white men who collected on the border of the Crow Creek reservation, ready to rush in and seize the Indian lands if the President proclaimed their opening.

Learning that the Crow Creek chiefs showed signs of panic, the land commission left Lower Brûlé and moved to Crow Creek,

[11] Secretary Teller did not produce a copy of this message when the Senate demanded all documents connected with the Sioux land commission's activities. Herbert Welsh obtained a copy at Lower Brûlé in the summer of 1883 and printed it in his pamphlet, *A Visit to the Sioux Reservation*, 12. Perhaps Agent Parkhurst gave Welsh the copy of the message. After uncovering the Captain Dougherty and Reverend Burt conspiracy and loyally aiding Shannon and Hinman in their attempt to pummel his Indians into submission, Parkhurst was being removed as agent, and he was highly annoyed with his Washington superiors.

where it started an intensive campaign to force the chiefs to sign. The commission's arrival proved disastrous to the frightened Indians, for it had hardly arrived when some miscreants (perhaps its own paid agents) started a rumor that President Arthur was on the point of signing a proclamation opening Crow Creek to settlement. The crowd of white men gathered on the southern border of the reservation—eager to have first choice of the Indian lands—now made a rush across the line and began staking out claims. They ordered the Sioux off and took possession of some of the Indian log cabins and farms. The honorable land commission, taking full advantage of the panic among the Indians, redoubled their pressure, and on the following day the chiefs were led to the table and their names set down on the agreement, which was then declared to be formally approved by the Crow Creek Sioux.[12]

The commission now announced that its labors had been successfully concluded, although not one Lower Brûlé chief had signed. In Washington, Secretary Teller approved the agreement as completed and in good order; President Arthur signed it and sent it to the Senate for final action. The document bore the names of only 384 Sioux, Newton Edmunds having persuaded Secretary Teller to adopt the Dakota legal view, that only the names of chiefs and headmen were required. The Dakota doctrine was a curious bit of legal thinking. It held that when the government ignored the treaty in 1876 and seized the Black Hills and other Sioux lands, forcing the chiefs and headmen to sign, it had established a precedent, and having violated the treaty once, it was free to do so whenever it chose.[13] Now, in 1883, the Indian

[12] This was another shady act. The agreement at Crow Creek was not the one signed at the other agencies, but a special one specifying that half the best Crow Creek lands should go to the whites and that the Crow Creek Indians, in return for being permitted to keep half their lands, should accept the Sioux land agreement. This was a shakedown, and in their eagerness to take more land from the Crow Creek Indians, the commission caused a legal doubt whether the whole Sioux agreement was not canceled because of the peculiar form of the document signed at Crow Creek.

[13] Some years ago the Sioux brought suit in the Court of Indian Claims for $882,457,354 for the seizure of the Black Hills and other lands in 1876, and the case has been decided in their favor. If the Edmunds land agreement had gone through, the Indians undoubtedly would have obtained many millions of dollars in additional compensation for the loss of lands under that agreement.

Rights Association and the other Indian welfare societies denounced this Dakota doctrine, declaring that the government's seizure of the Black Hills was a shameful act, to be shunned in future, and not a legal precedent to be followed. As the weeks passed the Senate remained ominously inactive, while the voices of the Indian Friends rose into an ever increasing roar of protest. Many men were shouting *stop thief*, and the sponsors of the Edmunds land agreement grew more and more worried. Then, on March 3, 1883, the blow fell, the dying Congress in one of its last acts approving a sundry civil appropriation bill bearing a rider which provided that the Sioux land agreement must be returned to the reservation for the assent of three-fourths of the adult Indian males. Thus the land agreement that had come into the world as a rider on a sundry civil bill was practically cast into oblivion through a similar rider.

For a man who had never been in Dakota, Secretary Teller exhibited astonishing eagerness to see the Edmunds land agreement put into effect. In that last-hour session of the Senate on March 3, he had President Arthur sitting in a room off the Senate chamber, waiting to sign the Sioux agreement the moment the Senate gave its approval.[14] When the Senate ordered the agreement taken back to the Sioux for the approval of three-fourths of the adult males, Teller with his faith still unbroken reappointed the same three men—Newton Edmunds, Judge Shannon, and James H. Teller—as the members of the new commission which was to resume the attempt to obtain the Sioux approval of the agreement. Newton Edmunds and Judge Shannon, who in September, 1882, had assured Secretary Teller that it was impossible to obtain the names of three-fourths of the adult male Sioux, now, in 1883, with the Sioux aroused and determined not to sign, adopted the attitude that the obtaining of thousands of Indian names was merely a clerical labor. Instead of going to the reser-

[14] This was told to a gleeful gathering of Indian Friends at the Lake Mohonk conference in October, 1886, by Senator Henry L. Dawes, who had led the last-minute fight to send the Edmunds agreement back to the Sioux. Dawes told the Indian welfare leaders at Lake Mohonk that Teller kept President Arthur in the anteroom until two in the morning in the hope that the Sioux bill would pass and could be signed.

vation themselves they sent their invaluable interpreter, the Reverend Samuel D. Hinman, to coax the Sioux into signing. In April, 1883, Hinman set out for Pine Ridge to begin his task.

Agent McGillycuddy had received orders to facilitate the work, and as soon as Hinman arrived, he called a council of chiefs. Red Cloud promptly got up and opposed the signing of the agreement; but after a violent quarrel as to whether this chief spoke for the tribe or only for himself and his blatherskite followers, Red Cloud accepted a dare and agreed that Hinman should go to the Indian camps and begin the experiment of obtaining signers. If a majority could be shown to be willing to sign, a great public meeting would be held and the men who wished to sign would do so before all the tribe.

Hinman now set out for the eastern part of Pine Ridge, where the progressive Bear People who were supposed to be strongly for the agreement had their camps. With him went Sword, the captain of police, and McGillycuddy ordered the policemen in each Indian camp to aid with the work. The name hunters first stopped at Two Lance's house on Wounded Knee Creek, and the next day the news ran through the reservation that little boys (some of them babies of three) were being permitted to hold the pen while their names were written down. This information was sent to some leaders of the Indian Rights Association in the East, and an uproar ensued. When publicly accused, Hinman stated that the Sioux were so eager to sign the land agreement that many of them wished their little boys' names also written down, so that in later years the boys could claim the honor of having signed this great document. He stated that his intention was to remove the boys' names from the final lists; but we have only his personal word for that. When called to testify, Sword and other policemen and the chiefs gave different versions of the transactions at the Two Lance cabin; but some of the most reliable witnesses stated that Hinman had himself asked the fathers to let their little boys sign the paper. Moreover, they stated that in his talk in the Sioux tongue the Reverend Hinman told them that only those who signed would be permitted to remain on the reservation. This seems to have been why Chief Little Wound told the Dawes committee that his band were eager to sign, as they did not wish to be removed to Indian Territory. However this may have been, Hinman continued for some time to take down the names of Sioux boys. At the little log school on Wounded Knee taught by the Sioux fullblood, Amos Ross, 18 out of 19 boy pupils signed the paper. In all Hinman put down the names of 144 boys, ranging in age from three to sixteen. Thus the merely clerical work the honorable commissioners had sent him to accomplish was started.

Testifying later before the Dawes committee, the Pine Ridge Indians accused Hinman of making threats to frighten men into signing the agreement, the threats ringing all the changes from having their rations cut off to removal to Indian Territory. They also testified that this man made free use of promises of various rewards to any man who would sign. The Sioux were passionate collectors of papers signed by white men in authority, promising them this or that. Thus Slow Bull was anxious to get possession of some good lands at the mouth of Wounded Knee Creek for his family or group of families, and Hinman wanted the names of Slow Bull and his friends for his land agreement; so presently Slow Bull was parading about displaying a paper signed by the Reverend Hinman which stated that the bearer had signed the agreement and wished to locate at the mouth of the creek. When faced with this matter later on, Hinman denied promising Slow Bull anything, and the careful wording of the paper Slow Bull held bore him out. Nevertheless, the Indians thought that Slow Bull had signed the agreement with the conviction that he was to be paid by being given the lands he desired. Hinman admitted signing many papers for the Indians, but denied any wrongdoing. It simply did not seem to occur to him that he was engaged in very important work and that it behooved him to walk carefully.

After the most severe labors, Hinman announced in August that he had the names of 633 Pine Ridge Indians who were ready to sign the agreement in open council, which he claimed was over three-fourths of the adult males. He gave the total of adult males as 969; but Agent McGillycuddy had been claiming a figure of 1,226 and for years had been obtaining funds from Congress to feed, clothe, and care for that number. To the agent and the

Reverend Hinman this seemed a small difference, and McGillycuddy acquiesced in calling a general council for the public signing of the agreement. But at this point the chiefs stepped in and stopped the proceedings. Red Cloud was away in Wyoming, visiting the Shoshoni tribe, and even his enemy Chief Little Wound refused to sanction a council in his absence, stating that it would not be fair to decide this very important land question while Red Cloud was away. Hinman therefore put off the final act and departed northward, to find out how the fishing was at Standing Rock.

Here a council was called, but despite Agent McLaughlin's willingness to obey his orders and assist Hinman in every way, the Standing Rock chiefs declined to have anything to do with that gentleman. They horrified him by stating that they had news that another commission (they meant the Dawes Senate committee) was coming to hold councils, and they did not wish to sign any more papers until they had seen these new men. McLaughlin told Hinman that his progressive Yanktonais Sioux were willing to sign, but were being intimidated by the arrogant nonprogressive Hunkpapas and Blackfoot Sioux. This was nice for Chief Grass, whose Blackfoot band had always been rated the most progressive at Standing Rock. They now disapproved of signing away their lands; therefore they were nonprogressives. A goodly number of Sioux from Cheyenne River Agency were present at this Standing Rock council, and they told the Reverend Hinman bluntly that he had better keep away from their agency, where the Sioux were very mad and none had any intention of touching that paper again.

Similar news reached Hinman from Crow Creek, where the chiefs now regretted their weakness in signing the agreement in February. Old Wizi (White Ghost), the head chief, was attributing the recent deaths of his children from whooping cough to his sin in signing the land agreement.

The Crow Creek and Lower Brûlé Sioux now had a new agent. This was John G. Gasmann, who had been agent to the Yanktons for six years and was known and liked by most of the Sioux. He

Courtesy Pitt Rivers Museum

Red Shirt, son of Red Dog

Courtesy Pitt Rivers Museum

Old-Man-Afraid-of-His-Horse

Courtesy C. I. Leedy

Crow Dog

had just taken charge at his new post when the Crow Creek chiefs pleaded with him to inform the Great Father that they wished to have the chiefs from all the Sioux tribes come to their agency for a great council, to decide what would be wisest to do in the matter of the tribal lands. Gasmann sent this seemingly harmless message to Washington, and almost lost his official head for doing so. He received by telegraph a curt order to do all in his power to prevent the Sioux from holding any such general council, as it was the policy of Secretary Teller to deal with each Sioux tribe separately. From what we have observed, this seems to have been an understatement of Teller's policy. It went further. It denied that the council of each tribe had any authority, that any chief had authority; it insisted that each Sioux man should act as an individual, so that he might the sooner succumb to cajolery or threats and do the bidding of the land commissioners.

The Reverend Hinman, taking a look all around, reported to his masters of the land commission that he seemed to be completely surrounded by Sioux roadblocks, with no hope of progress in any direction. The Indian opposition to the agreement was stiffening every day; the missionaries on the reservation and the Indian Friend groups in the East were shouting *stop thief;* the Indian Office had been frightened by a demand from the Indian Rights Association for full information concerning the Edmunds commission; and the Senate had alarmed Secretary Teller by a peremptory order for all documents, letters, and telegrams sent or received by this commission. Worse soon befell these earnest laborers in the Sioux vineyards. On March 2 the Senate passed a resolution appointing a select committee to "examine into the condition of the Crow and Sioux Indians, and to report on the advisability of the proposal to reduce the size of the Sioux reservation." Strangely, the men most interested in shaking down the Sioux for at least fifteen thousand square miles of land seemed to regard the appointment of this Senate committee as a matter that concerned them little. Senator Henry L. Dawes of Massachusetts, the chairman of the committee, was interested in Indian welfare; but he was a levelheaded businesslike person, not at all the kind of man to stir up unnecessary trouble, and, for some time after the Senate acted, neither Secretary Teller nor the Edmunds commission seemed in the least perturbed.

Then, in July, out of a blue sky, Secretary Teller received a polite letter from Senator Dawes, stating that before starting for Dakota he thought it only fair to inform him that he had heard from a very prominent man just returned from the Sioux reservation [Episcopal Bishop Hare] that the Sioux commission had failed to inform the Indians that under the agreement they were to give up a great tract of land, that the commission had threatened the Sioux at several agencies, and that the names of little children were being signed to the agreement. He also stated that, some time back, he and Senator Morgan had looked into the record of the Reverend Hinman's part in helping to frighten and force the little Ponca tribe to give up their lands in Nebraska and remove to Indian Territory, and that his and Senator Morgan's confidence in Hinman's integrity had been shaken. This message was a fair warning that Dawes was going to start shooting, and Secretary Teller sent a copy of the letter to Newton Edmunds at once, advising him to "look into the matter" as the Senate committee would "soon be in your section" inquiring into details. Edmunds wrote at once to all the Sioux agents, asking if his commission had threatened their Indians or tried to deceive them. All of the agents, except Parkhurst of Lower Brûlé (who had been removed) replied, some cautiously, others with breezy frankness, stating in effect that the commission had been fair in its dealings with the Sioux and had made no threats. Bishop Marty sent Edmunds a letter expressing his shocked astonishment that anyone should have accused the commission of misconduct in any form. Edmunds filed these letters, put his papers in shape for inspection, and waited with a quiet mind for the Dawes committee to come to his office in Yankton and hear his version of the matter; but the committee came into Dakota Territory by the back door, along the Northern Pacific, went straight to the agencies and began to take testimony from Indians, agents, employees, and army officers. On learning of this, the Edmunds commission threw up its hands in horror, and its peace of mind flew out of the window. The

Dakota land boomers, becoming suddenly highly agitated, turned in a general alarm.

Meantime Secretary Teller had written to urge the Dawes committee to see the Edmunds commission and get its views before going to the reservation. "I suppose you are aware," he wrote, "that a great number of people in that vicinity do not want the Indians to part with any portion of their land, and in fact they would much rather have them remain wild Indians than have them become civilized; and in some cases those who profess the most love for the Indians are the ones who for selfish purposes prefer the Indians to remain Indians." And who were these Dakota bad men? It seems incredible, but the official records and the testimony taken by the Dawes committee makes it certain that the villains thus glanced at by Secretary Teller were Bishop William H. Hare and Bishop H. B. Whipple of the Episcopal church, the Reverend T. L. Riggs, Congregationalist missionary at Cheyenne River, the Reverend W. J. Cleveland, Episcopal missionary at Rosebud, the Reverend H. Burt, friend-in-need to the frightened Crow Creek and Lower Brûlé Sioux, and the Reverend Luke Walker, fullblood Sioux, missionary and teacher at Lower Brûlé. These were the men Secretary Teller was accusing of working to keep the Sioux in a state of barbarism, and of opposing the land agreement for their own selfish purposes. Never was a baser charge brought against men who were devoting their lives to the service of others. Every one of these men approved of the sale of the surplus Indian lands. What they did not approve of was cheating the Indians or obtaining their lands by threats. Bishop Hare was already advocating the sending of a new land commission to the Sioux with an offer of fair terms, and he promised the eager support of all his missionaries to such a project. But Secretary Teller for some reason was wedded to the Edmunds commission, and he turned a deaf ear to any suggestion for change.[15]

The Reverend Hinman had been sent back to Pine Ridge, perhaps in the hope of making some showing by obtaining the approval of the Sioux at one agency for the Edmunds agreement. On

[15] The Dawes, Teller, and Edmunds letters are printed in the *Dawes Report*, 193, 374–75.

reaching the agency, he apparently listened to Agent McGillycuddy alone, for his report has much the sound of one of that agent's denunciations of Red Cloud. Hinman reported that on his arrival he found all of his high hopes blasted by the action of this chief, who was engaged in proclaiming "a barbarian empire" among the Sioux, with himself as the emperor. In simpler language, the Sioux at the various agencies, worried by the actions of the Edmunds commission, had asked Red Cloud to take the lead in an effort to save their lands. Hinman went on to state that any Sioux who signed the agreement would be banished by the tribe, and that Red Cloud was sending runners to every camp on the reservation, urging the Indians to refuse to sign any more papers. Hinman took one good look around, threw in his hand, and left the game. That was the end of the Edmunds land agreement. The Dawes committee, after visiting every agency and taking the testimony of hundreds of Sioux and whites went on to Omaha, where at last they gave the Edmunds commission an opportunity to present its case. Edmunds and the others made a general denial of all charges and then introduced a Dakota orator who favored the Senators with a stump speech in which he attempted to demonstrate that the Edmunds agreement had been properly approved by the Sioux nation and should be passed by the Senate. The Senators listened politely, but were not impressed.

Thus the Dakota leaders, aided at every point by the powerful Secretary of the Interior, had failed in their effort to take the Sioux lands; they had failed, and they had produced a condition that would render success of any future land agreement extremely doubtful. The Edmunds commission had come among Sioux Indians who were friendly and most of them trusting; but the treatment they had received at the hands of the commission had united these Indians in bitter hostility to the land deal. The Sioux no longer trusted white men, even when they came from Washington pretending to bring the words of the Great Father. At every agency the Indians were now watching, angry and suspicious, for the next move of the whites.

5

The Brethren

FOR A NUMBER of years prior to 1882 some of the more intelligent Sioux had realized the fact—very strange to their way of thinking—that their people had many faraway friends among the whites: men, and women too, who had never seen a Sioux but who nevertheless felt that they were friends of the tribe, and good friends, ready to extend a helping hand when it was needed. To the average Sioux, who had not a trace of altruism in his makeup, this fact seemed contrary to simple reason, and for a long time it was not credited. A Sioux had to see a man daily and learn to know him very well before he began to feel any friendship for that man, and as for helping others, the Sioux saved such special favors for his own close kindred. Thus for some years the people on the great Sioux reservation found it difficult to believe that they had unknown friends among the whites; but the effective work these white friends performed in 1882–83 in blocking the Edmunds commission's effort to take their land at last convinced some of the Sioux.

But just who were these white friends to whom the Sioux chiefs were now turning in the hope of assistance? They were men and women of a superior type scattered through every eastern and northern state of the Union. They were all devout Christians, high-minded people whose motives the average Sioux could not possibly comprehend, but whose good intentions would have to be taken on trust. In New England people of this sort had begun to interest themselves in Indian welfare work even in Colonial times, and they had formed the American Society for Promoting Civilization and General Improvement of the Indian Tribes of the United States at New Haven, Connecticut, about the year 1820. Other associations with similar lofty names and purposes had been formed in other New England communities. All of these groups had close church connections, always Protestant; and the Quakers must not be overlooked, for they were in the Indian welfare movement from the very start. From about 1840 these little societies of Indian Friends had had their attention diverted by the antislavery movement, into which they flung themselves with burning passion. When the Civil War came, they sent pressure groups to Washington to agitate for the immediate freeing of the slaves, and when that act was accomplished, they took up the new task of caring for the freed Negroes. It was at this period in their activities that they learned, with the aid of such men as Thaddeus Stevens, the art of obtaining by unified effort large sums of money from the government; and now, instead of contributing their own dollar or fraction of a dollar to their favorite cause, they put pressure on Congress and reached into the treasury for millions, to be spent as they dictated on plans of their own making.

It was in 1865 with the Civil War's ending that these Eastern benevolent societies had their attention drawn once more to the needs of the Indian tribes, and although their main interests still lay with the Southern Negroes, they took up the cause of the Indians with such zeal that within a year they were strongly influencing the officials in Washington. First and always for peace, they threw their weight back of the Indian peace treaties of 1865, which brought no peace, and they were enthusiastically in favor of feeding and caring for the wild Plains Indians—a form of bribery that was supposed to content these Indians and insure lasting peace. But, despite the immense sums spent on this new ration system, the southern Plains Indians went to war in 1874 and the Sioux war followed in 1876–77. The Christian Peace Policy had failed, and the Grant administration turned a deaf ear to the pleading of the Christian Indian groups and resorted to force in dealing with the wild tribes.

With the Sioux war over, Grant gone, and a friendly new President—Rutherford B. Hayes—in the White House, the brethren took heart again and came forward with a new Indian policy, which was simply their old and very much smashed-up Indian Peace Policy rebuilt on a slightly revised plan. One of its main features now in 1878, as in the days of 1868, was a scheme for turning all the Sioux rapidly into Christian farmers; but the leaders of the Christian and benevolent groups had learned a little in the past ten years, and they no longer claimed that this work among the Sioux could be accomplished in four years' time. They were careful now not to specify a certain period, but they did express the strongest conviction that this work could be accomplished speedily. Their policy was adopted, and beginning in 1878 fresh efforts and great spending of public money went into the scheme for turning the Sioux into self-supporting Christian farmers. Yet after five years, in 1883, one of their own friends, Senator Dawes of Massachusetts, made a most thorough examination into conditions among the Sioux and reported that very few of these Indians were raising enough food to supply their own families, while the great majority were simply not interested in the matter of self-support. The leaders among the brethren, reading that blasting report, were undismayed. They never gave up. They had been proved wrong about the Sioux desire to farm, so they shifted their ground a bit and began to talk of methods for forcing these Indians to work. Of their continued authority to participate in Indian affairs they had no doubt at all, and they regarded their volunteer efforts as of vital importance if the country was to have an honest and effective Indian service. Like all doctrinaires, they believed firmly that they were indispensable, and they never admitted a mistake. When accused from every quarter, after 1868, of blundering and slackness, they invented a mythical Indian Ring on which to saddle the blame for their own shortcomings. It was the era of rings. The public had been horrified by the evil deeds of the Whiskey Ring, the Star Route Ring, the Tammany Ring, the Railroad Ring, and many other circular groups of rascals. So why not also an Indian Ring? The brethren made one up, nominating the members to suit their whim or spleen of the moment. Some members lived on the reservations, but the principal leaders and plotters of the ring were Indian contractors, who carried on dark intrigues in Washington. In fact, these men did just what the immaculate brethren were doing—they pulled wires in official circles and put pressure on Congress. But the ring members worked for their personal gain, whereas the brethren labored in a noble cause. Over a period of two decades no real evidence of the existence of such a ring was produced. There were plenty of rascals engaged in plundering the government and the Indians, but they were either lone wolves or their organizations were purely local. Agent McGillycuddy of Pine Ridge had his own private Indian Ring in which he included anyone he took a dislike to. In his spite he even nominated some of the brethren to this round table of rascals.

The very costly system for feeding, clothing, and caring for the Indians which these benevolent societies had supported and helped to build up was the direct cause of most of the dishonesty of Indian agents and contractors after 1867; but the Indian welfare groups never lost faith in their own ability to deal with any situation. They developed a plan for a new Board of Indian Commissioners, which was to guard all Indian funds and keep a watch on their expenditure. This board was to consist of "ten persons eminent for intelligence and philanthropy." And where in all the land were ten such persons to be found except among the leaders of these same Indian welfare groups? This plan was brought before Congress in 1869, and as soon as it became law, President Grant did as he was expected to do, nominating a full board chosen from the leadership of the Christian and benevolent societies.

These high priests of the Indian welfare groups had hardly had time to organize their board when their chairman, William Welch, a wealthy Philadelphia merchant, resigned when Commissioner of Indian Affairs E. S. Parker stood on his legal rights and refused to turn over control of all Indian funds to the new board. It then developed that the law establishing this board had left its powers up in the air and that it had no authority except to make investigations and to recommend action. However, the board members could travel to the ends of the country at public expense, could snoop endlessly into the affairs of any person connected with the Indian Service, and could then print their findings at public expense. This they proceeded to do with the loud reverberations that are the accompaniment of such investigations. The board's first victim was the man who had first dared to outface them: Indian Commissioner Parker. They charged him with wrong-doing and pulled him down. Then they turned their attention to bigger game, the Secretary of the Interior, Columbus Delano. Mr. Delano, however, fought back with such vigor that he forced the resignation of the board's obstreperous secretary, Vincent Colyer.

When the Grant administration turned its back on the Indian Peace Policy during the Sioux war of 1876, the leaders of the Indian welfare groups who made up the membership of this Board of Indian Commissioners were left high and dry. In the hurly-burly of an Indian war no one listened to their views, and presently the six leading members of the board resigned in a body and walked off the stage. This ended for a time the attempt of the Indian welfare societies to control the government policy, but the crusaders never lost hope. They waited for the Sioux war to end, in the meantime working to improve their position.

By 1880 the roll of their groups was impressive. The list included the Indian Rights Association, the Indian Humanitarian Association (founded by Herbert Welsh of Philadelphia), the Indian Defense Association, the Boston Indian Citizenship Committee, the Women's National Indian Association, the Indian Hope Association, and the Indian Treaty-Keeping & Protective Association. Besides these, there were numerous smaller groups.

The operations of these societies are an interesting study. Their leaders held a conference every summer or fall, a closed meeting at which Indian policy and the program for the coming session of Congress were discussed. In early winter, when Congress assembled, a larger group of Indian Friends met in Washington in the parlor of the Riggs House hotel, a meeting which was public but carefully arranged so that the program already decided on by the leaders of the Indian movement would be presented in the most favorable light. Guests who did not exhibit a proper spirit of co-operation at one of these public meetings were not encouraged to come again. Dr. T. A. Bland, a retired minister who held views concerning Indian policy that did not square with the opinions of the high priests, was thus gently shoved out of the sheepfold in the Riggs House parlor. He promptly organized an Indian association of his own, founded an Indian paper in Washington, the *Council Fire*, and started a vigorous opposition to the Indian policy advocated by the orthodox groups. Moreover, he collected Red Cloud, Spotted Tail, and some other Sioux chiefs, made friends with them, visited them on the reservation, and thus obtained first hand information as to what the Indians' own views were. Here Dr. Bland had something that the leaders of the other Indian welfare groups did not have, and probably did not desire. Their Indian policy was not based on consideration as to what the Indians desired but on their own views as to what would be best for the Indians. Their policies were symbolized by the fact that they never contacted a Sioux chief without at once trying to persuade him to don a stiff-bosomed white shirt, a black suit, and a shiny silk hat. At the root of all their high-sounding policy talk lay this earnest desire, to make the Indians over into as perfect imitations as possible of their own perfect selves.

A typical January meeting in the parlor of the Riggs House was opened with prayer followed by the singing of hymns. The various Indian mission groups of the Protestant churches then gave an account of their activities during the year. A guest of honor then addressed the assemblage. The leaders always tried to get the chairman of the Senate or House Committee on Indian Affairs to make the opening speech. Occasionally they induced the Commissioner of Indian Affairs to talk, and on a few rare occasions the august Secretary of the Interior favored one of these meetings by being present. The tone of these meetings was strongly religious, and often the air was as tense with emotion as at a revival. Nothing delighted the assembly more than to sit and cry gently while listening to an Indian boy singing familiar hymns in the Kiowa or Pawnee language.

No reasonable person could have possibly objected to any of this. It was when the leaders of the Indian societies arose and presented the program of Indian policy for the coming year that it was time for someone with a sense of proportions to object. This

Indian policy, the work of a little closed group, was put forward as the best possible government policy for the coming year; the meeting then was asked to approve, which it nearly always did by unanimous vote, and the machinery of pressure on Congress and the administration was set in motion to force a privately concocted Indian policy through Congress and make it the law of the land.

Most of the officials of the government Indian Office and many of the members of Congress were afraid of these well-organized Indian Friend groups. They maintained a paid permanent secretary with an office in Washington, Professor C. C. Pointer, a pioneer of the modern school of lobbying. He knew all of the influential officials and members of Congress and made use of them to insure the passage of bills the Indian societies were backing. He fought outlaw Indian legislation, which included almost any Indian bill sponsored by the ungodly or by Indian welfare organizations not included in his own orthodox groups. If Congress exhibited an unwillingness to enact the bills Pointer was working for, he sounded a general alarm by printing leaflets and posters to be distributed in churches or town halls, urging all good Christians and Indian Friends to write at once to their members of Congress and demand action.

In 1883, a Quaker, A. K. Smiley, who was a member of the Board of Indian Commissioners, invited his colleagues on the board to spend several days as his guests at the Mountain House, a hotel at Lake Mohonk in New York state. Smiley and his wife, who owned and managed the hotel, provided every comfort for their visitors and fed them lavishly. The whole gathering talked Indian welfare and Indian policy day and night, and they all enjoyed the stay at the Mountain House so much that Smiley suggested such conferences should be made an annual event. Thus the famous Lake Mohonk conferences were inaugurated, and each year thereafter the Smileys entertained a group of carefully chosen guests for several days at the Mountain House, usually in the autumn. At these meetings men and women, most of whom had never been in real contact with Indians and who were amazingly uninformed of Indian ways of thought and action, outlined a government Indian policy, which was taken to Washington and put through by pressure methods.[1]

By such methods a small group of fervent crusaders built up the so-called Indian problem from nothing in 1864 to the greatest issue before the American nation in 1884. Senator Henry L. Dawes of Massachusetts, who often worked in friendly union with these Indian societies, set one of their meetings to purring in 1885 by telling them in an address that they, by their endless labor, had made the Indian problem the great issue of the time. This was true; but the great issue was one which the mass of the public took almost no interest in; it was an artificial affair, a toy balloon blown up until for the moment it overshadowed every other matter dealt with in Washington.

It would be very unfair to say that these Eastern friends of the Indians did not accomplish much good by their strenuous efforts. Yet as a rule when their plans benefitted the Indians in any way it was more by accident than by design, for these people did not understand Indians, and usually they were quite incapable of judging what would help the Indian and what would injure him. These good men and women, indeed, mistook their lively interest in the Indian for deep understanding and imagined that they were holding out hands in help when usually they were only meddling in matters beyond their range of comprehension. "The Indian tribe," they said, "dwarfs and blights the family; the tribal system paralyzes labor; it prevents all accumulation of property for the children; it cuts the nerve of individual effort." And Congress, impressed by this high-sounding nonsense and pressed for action, passed a bill outlawing chiefs and tribes. To do this, Congress violated a score of treaties which it had no authority to touch, and let the Indians in for a ten-year period of chief-baiting, featured by seven years of uproar produced by Red Cloud and Agent McGillycuddy. Throughout this crusade against the tribes and chiefs, every Sioux agent and many of the officials in Washington knew that the average Sioux could not live outside his tribe,

[1] For the Lake Mohonk arrangement, see L. B. Priest, *Uncle Sam's Stepchildren* (New Brunswick, N. J., 1942), 84.

that he could not think for himself, and that (come what may) he would always follow a chief. McGillycuddy knew that. While publicly and vociferously trying to pull down Red Cloud he was at work building up other chiefs, whom he hoped to dominate.

When Cleveland was elected in 1884 it was a shock to the Indian welfare leaders, who had small faith in the possible virtue of a Democratic administration. They were not reassured when the new President appointed as secretary of the interior, in charge of Indian matters, L. Q. C. Lamar of Mississippi, a former Confederate officer and a former slave owner. However, the welfare leaders considered it their duty to try to win over this man to the acceptance of the views on Indian policy that they had been building up for years, and they sent a delegation to Washington to see Lamar. These men entered the secretary's office and began to instruct him. "The Indian tribe," they said, "dwarfs and blights the family; the tribal system paralyzes labor; it prevents . . . ." Lamar lifted a deprecating hand. "Gentlemen," he said, with a faith as stiff as their own, "the tribe is the Indian's natural political condition, and he should by no means be removed hastily from that condition until he is fitted for a higher one." And that was the end of tribe busting and chief baiting as a government policy.

These Indian welfare groups could never comprehend that the Indians were of a separate race, differing from themselves, mentally, morally, and physically. Their prime object always was to make the Indians over into perfect replicas of themselves. *They* were energetic and resourceful; if given free farming tools *they* would go to work and make a success of farming; therefore the Indian would. *They* with as much government aid as was now provided for the Sioux would be on their feet and prospering in just a few years' time; therefore the Sioux should do the same. *They* realized the advantages of education; so the Indian must also be eager to send his children to school. None of this was true. The facts in every instance glaringly contradicted these doctrines, but the welfarers had little use for facts, except those that fitted in with their preconceived notions. Faith was better than facts. From the beginning to the end of their great crusade they exhibited no understanding of the Indians. They could not even understand the white men who lived with the Indians. These so-called squawmen they insisted on regarding as disreputable and lazy scoundrels who had gone to the reservations and taken to living with Indian women so that they might obtain free rations, clothing, and other needs. Among the Sioux there were some white men of the type described; but the greater number were former Indian traders whose business had been destroyed when the Sioux occupation of buffalo hunting was ended and the Indians went to the reservation. Having ruined the traders by stopping Sioux self-support through hunting, the government made use of these men, who spoke the Sioux language and had much influence in the tribe. It employed them to coax the Sioux to the reservation, and the officials made great promises of what they would do for these whites if they would stay on the reservation and aid in starting the Sioux on the way toward civilization. Most of these squawmen helped the government loyally. They started farms, and if the Sioux did not follow their example it was not their fault but due to the ingrained opposition of the Indians to any form of manual labor.

The officials soon forgot the promises they had made to the squawmen, and presently the leaders of the Indian welfare groups began a campaign of detraction against these men as a class. The older squawmen had married into the tribe years before any free rations or free clothing had been thought of, at a period when there was not one Christian minister within hundreds of miles of the Sioux camps. They married by tribal custom; and now men with narrow minds could say that they were living in sin, setting the Indians a bad moral example. Frequent demands for their removal from the reservations were made, but there was no legal ground for action. The white men were members of the Sioux tribe, most of them with grown children who were also legally Sioux. But the pressure of the Easterners did induce some of the Sioux agents to persecute this class of whites. McGillycuddy outlawed all of them at Pine Ridge, and for seven years he listed them as undesirable persons who were illegally present on the reservation.

Much of the talk concerning the wicked deeds of the squawmen and the bad influence they had on the Sioux had its origin in the fact that these men read the newspapers and kept the chiefs informed as to what was going on in Dakota and in Washington. In the view of some of the agents this was a serious offense, and the whites and mixed bloods who could read were accused of stirring up trouble. The most notable example of this was when they told the Sioux in 1882 the truth they had read in the Dakota newspapers concerning the Edmunds land agreement. Not until the squawmen and mixed bloods told them did the Indians understand that the paper they were being urged to sign would deprive them of half their lands.

The Dawes Senate committee in its extensive investigation on the Sioux reservation in 1883 learned much of the truth concerning the squawmen and mixed bloods. Whenever they saw a little farm with fields fenced and crops well cared for they almost always found that it belonged to a squawman or to a mixed blood. They found most of these men willing at any time to help without pay in any attempt to improve the condition of the Indians. In 1890, at the Lake Mohonk conference, the Reverend T. L. Riggs, Congregationalist missionary to the Cheyenne River Sioux, told the assembled brethren to their faces that the squawmen and mixed bloods, whom it was their fashion to denounce as undesirables and men of disreputable character, were always the first to come to the aid of the missionary or teacher when anything had to be done to help the Indians, that the homes of such men were better built and cared for than any belonging to full bloods, and that this class of squawmen and mixed bloods always set a good example for the Indians to follow.[2] Theodore Roosevelt when he visited the Sioux reservation as a civil service commissioner in 1892 wrote: "I became satisfied that on the whole the half-breeds are very much farther advanced than the Indians proper, and that in a great number of cases they are on a complete level with the whites; while the squawmen in a great number of instances are of great benefit to the Indians, serving as connecting links between them and civilization and rendering the road upward much easier for them . . . . Indeed, I became satisfied that

[2] *Report of the Board of Indian Commissioners* (1890), 145.

the attacks on the half-breeds and squawmen as classes were entirely unjustified."[3]

The inability of many Easterners to understand the Indians, mixed bloods, and squawmen, linked with their fervent purpose to meddle in the lives of these people produced both tragic and comic effects. Orange Judd, a famous man and publisher of farm books in the eighties, well-to-do, plump and rosy, once told a highly appreciative gathering of ladies and gentlemen how he *loved* his Sioux brothers (whom he had never met), how he *longed* to put on a blanket and go and live the simple and frugal life of these children of nature. If he had done so! We may picture him departing in confusion from the reservation and heading swiftly for home, feeling anxiously in his mouth with one finger as he sped on his way, for the grim Sioux diet of just tough beef and soggy bread made in a skillet was in itself enough to rout the most sentimental of white visitors. This diet produced scurvy in whites unaccustomed to it, and if a man did not take prompt action his teeth fell out.

The brethren were deeply grieved over the hostile attitude of the frontier population toward their Indian brothers, and after thinking the matter over for several years they produced the usual perfect plan. It was a scheme for moving the wild Sioux, Comanches, and other tribes from the plains and resettling them in Eastern communities, where the civilized and kindly whites would love to give their Indian friends a welcome and aid them in a rapid advance toward civilization. These curious people discussed this scheme with great seriousness, developed elaborate blueprints, showing just how many Sioux were to be colonized in Ohio, how many Kiowas and Comanches in Indiana. They laid this plan before President Grant, and Grant let it lie untouched. Five years later the brethren, who never gave up, laid the same plan before their comrade, Carl Schurz, secretary of the interior, but even the impractical Schurz could see that this plan was moonshine. Nothing could induce people in Eastern states to permit the colonization of masses of wild Indians in their vicinity, and nothing short of war would force the Sioux and

[3] Theodore Roosevelt, *Report to U. S. Civil Service Commission* (1892), 9.

Courtesy Bureau of American Ethnology

Kicking Bear

Courtesy Bureau of American Ethnology

Ghost Dance Preacher

Sioux Ghost Shirt

Courtesy Smithsonian Institution

other tribes to submit to being moved into the country east of the Mississippi.

If the Sioux and other Western tribes had realized just what the intentions of the Indian Friend groups were they would have lost all confidence in them at once. At the very time in 1883 when the Sioux were feeling grateful toward their white friends in the East who had helped them in their effort to save their lands, the leaders of the Indian welfare groups were earnestly discussing a plan to take the lands from the Sioux—not in the wicked manner which Newton Edmunds and his comrades had adopted but by a model agreement, and this was to be done for the good of the Sioux as that good had been thought out by the leaders of the Indian Friends. Indeed, a whole new program was being planned for the benefit of the Sioux and other tribes. At any rate, the planners imagined that the Indians would be benefited. Thus the Sioux were to be deprived of half their lands for their own good. The Sioux *must* become farmers and support themselves, and (according to the high thinkers of the Indian welfare groups) this could never be accomplished as long as the Sioux had that immense reservation to wander about in. The Sioux had always lived in moving camps; and here on the reservation, just when their agent imagined that he had them all pinned down on farms and ready to work, they would betake themselves to their canvas tipis and go off on an entire summer's leisurely wandering, visiting relatives and friends at other Sioux agencies. The new plan was to give each Sioux family a farm and to sell the rest of the reservation lands to the whites. Thus, pinned down and wedged in among white settlers, the Sioux could not go wandering. They would have to remain at home and farm. That was the theory. People who knew the Sioux had very serious doubts about how the scheme would work out.

The leaders of the Indian welfare societies had now hit upon two new methods for swiftly advancing the Western tribes—education and citizenship—and they were preparing to launch a grand assault against barbarism and drive it from the Indian reservations. They had learned a little since 1868, when they had imagined that the Sioux would take eagerly to farming and schools. They now knew that most Sioux hated farming and that schools were far from popular on the reservation; therefore all the new plans for helping these Indians were carefully equipped with means for putting pressure on the Sioux. Thus they were to be forced to stay on farms, and if they still would not farm they were to be deprived of rations and starved into working. If they did not send their children to school armed Indian police would come and take the children and, if considered necessary, would put the father in jail. Another plan provided that families that did not keep their children in school should be deprived of rations. Indeed, the kindly members of the Indian Friend groups, who had probably never gone without a meal in their comfortable lives, seemed to think that whenever a Sioux or his family misbehaved in any manner the remedy was at hand—just cut off the rations and starve the family until it submitted. As a justification of their plans to starve Indian families these worthy people quoted St. Paul and the New Testament.

Perfectly assured of their right to meddle in the lives of the Sioux and other Indians, the leaders of the benevolent societies had only one doubt in 1884. They feared that they could not induce the new Democratic administration and Congress to co-operate in their Indian policy. For that matter, the outgoing Republican administration had not been particularly friendly, and Secretary Teller seemed to be furious over the part the Indian Friend groups had played in defeating the Edmunds land agreement. In the last days of his administration President Arthur took a final slap at the Sioux by issuing a Presidential order opening Crow Creek reservation to settlement. The leaders of the Indian Friend groups lifted their voices in horror at this brutal action. They rushed to Washington, and to their surprise and pleasure the new Democratic President readily consented to take immediate action to save the Crow Creek lands for the Sioux. Coming away from their conference at the White House, the Indian welfare leaders felt considerably reassured. President Cleveland had not only agreed in their view that the Indian problem was the greatest issue before the American public, he had stated that he would rather be known in history as the man who had solved that vexed question than to have any other title of honor. He also seemed very sound on civil service, allaying the fear that the Democrats would remove efficient Republican Indian agents and replace them by the appointment of worthy Democrats. And yet, it was Cleveland's appointments that were soon causing the leading Indian Friends to shake their heads and tighten their lips. As secretary of the interior, in general control of Indians, he appointed Lucius Quintius Cincinnatus Lamar of Mississippi, an hereditary slave holder and a leader in the late rebellion, a man who in the view of the Eastern church people and social reformers had led a bad life and who still was as unreconstructed and wrongheaded as ever. The Indian Office was handed over to two former rebels, Democratic wheel horses from Tennessee, A. D. C. Atkins, a one-armed Confederate colonel, becoming commissioner of Indian affairs, while the all important place of assistant commissioner in charge of personnel went to one Upshaw (he did not appear to have any further name) who seemed to be simply a Democrat whose waking hours were devoted to questions of party organization and strength. The welfare leaders set a watch on Mr. Upshaw.

Commissioner Atkins now surprised them by coming out strongly for their favorite plan of making all Indians into farmers. He authorized the appointment of many additional white farmers, to teach the Indians, and the brethren began to think better of him. Then cries arose from the Sioux reservation, where it was reported that Assistant Commissioner Upshaw had sent out to Dakota a small army of southern Democrats, some of whom no doubt were slightly acquainted with farming methods, but the majority seemed to have done all their plowing in political fields.

The leading members of the welfare movement, convening at Smiley's Mountain House at Lake Mohonk for the usual autumnal meeting in October, 1885, were still shaking their heads over the doings of Atkins and Upshaw at the Indian Office, but they were most determined to continue their efforts in behalf of the Indians. Let the Democrats in power do what they might to thwart them, they would go straight forward with their great crusade. The little inner group of leaders now produced a nine-point plan for disposing of all surplus Indian lands, after giving each Indian family a farm, for forcing the Indians to work and doing away with free rations, for giving the tribesmen citizenship at the earliest possible moment, and for offering all Indians that greatest boon of all, education, and seeing to it that they took it whether they liked it or not. Some men who were present expressed a fear that too much pressure in attempting to force the Indians along rapidly might prove a mistaken policy, but the majority of the meeting urged speed. The time was ripe, they said, for crushing all *Indianism* out of Indians by one great assault. Dr. Lyman Abbott and Professor C. C. Painter (he was the paid lobbyist for the Indian welfare groups in Washington) spoke fervently of uprooting the entire reservation system at once. Captain Pratt of Carlisle School went even further, advocating turning the Indians out into the mass of the white population and leaving them to sink or swim. Someone suggested that the Indians were not fitted to cope with civilization at close quarters and that they would surely sink. Captain Pratt cheerfully agreed to this view but did not seem to think it weakened his plan. A few Indians would survive, those who deserved to survive, he stated. These self-styled friends of the Indians sat about in comfortable chairs and discussed cutting off rations and starving Indians into going to work in earnest. They approved the plan, supporting their approval by quoting words attributed to St. Paul: *He who will not work shall not eat.* Did they have the faintest glimmering as to what it meant, in practice, to set about starving Indian families by deliberate plan?

Senator Dawes, listening to this talk, must have felt very queer. These men and women had not a doubt about their moral right to take from the Indian his land and the very food that kept him alive, once they had made up their minds that such action would be for the Indian's good. They particularly exasperated Dawes with their eager discussion of plans for destroying Indian treaties. He reminded them that a few years back when men outside their own groups had attacked the validity of certain Indian treaties they had sent the women members of their groups about the country for three seasons (1879–81) to collect signatures to a great petition which urged Congress to defend the treaties against all attacks.[4] These treaties he told them were as sacred today as they had been then, and under the Sioux treaty the lands of the reservation had been *sold* to that tribe by our government, these Indians having a title to their lands as good as any man present had to his own property. The Sioux treaty title, he went on, "is a title-deed as perfect as yours, and to talk of taking that land from them without their consent for their good is the same as talking of taking away our neighbor's title to his home for his good."[5] But Senator Dawes was wasting his breath. Like most crusaders, these leaders of the Indian welfare societies had the curious ability to hold high a certain matter one day, proclaiming its sanctity, only to cast it down and tread it into the mire the moment it stood in the way of some new plan of their own concoction. Dawes merely irritated them by reminding them of their past pledges and by pointing at fundamental laws that stood in the way of their newest schemes.

Having decided among themselves at Lake Mohonk on a nine-point Indian program, these men went to Washington and on November 10, 1885, made their wishes known to the new Secretary of the Interior. Secretary Lamar, former slave-owner and former rebel, was not a man they could like or respect; but they felt it to be their duty to attempt to lead him aright. They had made themselves the heads of the Indian movement and had been accepted as such by the three Republican Presidents who had preceded Cleveland; they had a definite program, and they seem to have hoped that Lamar, like the Republican secretaries of the interior, if firmly approached would adopt their policy as his own. They marched into his office therefore, strong in the faith of their leadership and good intentions, and set before him their program for disregarding the treaties, forcing land in severalty, destroying the reservation system, and making the Indians by one swift act

[4] The Women's National Indian Association conducted this treaty crusade. "This valiant and glorious band of women," as their friends termed them, obtained 100,000 signers to their petition and took it with solemn ceremony to Congress "wrapped in the glorious folds of our Nation's Flag." But now, in 1885, the Indian treaties stood in the way of these same crusaders and they quickly convinced themselves that it was a righteous deed to destroy the treaties.

[5] Speech of Senator Dawes and other speeches at Lake Mohonk conference, in *Report of the Board of Indian Commissioners* (1884–85).

citizens of the United States, equal in standing with all other citizens.

Secretary Lamar listened to them. He then told them bluntly that he would have nothing to do with any scheme for uprooting the Indians, in their present transition state, and suddenly tossing them into the condition of land owners and citizens, a condition which the great majority of them were in no manner fitted to assume. This former slave owner and Confederate officer was no stranger to causes and crusades. He had been in one himself from 1850 to 1865; he had been a leader in that Southern crusade to maintain slavery and to extend it to the territories, and he now said to these fervent Indian crusaders, "I am a conservative, and I have been made so by a costly experience." He stated emphatically that no hard and fast policy could be applied to all Indian tribes, for there were many varying states of development, while in general the average Indian was so backward that making him suddenly a land owner and a citizen would destroy him. It would be almost as cruel as a war of extermination. The Indians, he believed, should for the present be kept under reservation influences, and the tribal governments should not be entirely broken up. As for citizenship, Secretary Lamar stated that after swallowing four million illiterate former slaves as new citizens the nation would not have to strain much over a few thousand Indian citizens; but that it would be a sad day for the tribesmen when they were suddenly catapulted into this strange new status. It would be the end of most of them. Many intelligent Indians knew this and were in daily dread of being saddled with the taxation and other responsibilities that citizenship involved. The more advanced Indians he, Lamar, would lead toward land ownership and even citizenship; the backward ones he would make no attempt to rush along. Looking straight into the eyes of the Indian welfare leaders this man told them, "Those that are ready I will push on; those that are not I will protect." The welfare leaders said good day to Lamar with meekness and withdrew to plan their attack. There could be only war between them and this man. He had said "to take the Indian out of his tribal state is to change his social condition, his religious and hereditary impressions, before he is fitted for the higher civilization." Lamar openly defended *Indianism*, which they were determined to fight and to destroy, root and branch![6]

Their failure to dispose of their nine-point Indian program to Secretary Lamar was a disappointment, but it only stiffened the determination of the Indian welfare leaders to push straight on. For years they had been accustomed to coming to Washington in the autumn, obtaining the easy assent of the Secretary of the Interior to their plans, and then going through the farce of having these plans sanctioned by a meeting held in the parlor of the Riggs House. This time the Secretary of the Interior had rejected their program; yet, in January, 1886, they met in the Riggs parlor, and after listening to many fervent speeches from the welfare leaders they triumphantly adopted their nine-point plan. Posing as public opinion personified, they demanded the attention of the Cleveland administration. With a new battle cry that would twang the heart strings of every Christian, every idealist and educator in the land, they prepared to advance.

[6] The Indian welfare leaders' own account of this conference with Secretary Lamar is printed in *Report of the Board of Indian Commissioners* (1885), 115–16. See also Priest, *Uncle Sam's Stepchildren*, 238, 284.

6

A Very Little Progress

ON THE GREAT Sioux reservation the Indians heard few echoes of the doings of their Eastern "friends," who were preparing to dispose of the Sioux lands, to force the Indians to work, and, if necessary, to starve them for their own good. Alarmed by the operations of the Edmunds land commission of 1882, the Sioux were on the watch for any further attempts to take their land. Red Cloud, posing as the head of a new federation of Sioux tribes, was busy talking about united action to protect the tribal lands, and in this work he was being strongly supported by Dr. Bland of Washington, whose Indian Defense Association was fighting the program put out by the orthodox Indian Friend societies. Bland was also using his magazine, *The Council Fire*, to oppose sale of Indian lands, citizenship, breaking up of the tribes, and disregard of treaties, all of which were important features in the program of the Lake Mohonk group of Indian societies. Red Cloud—deposed, jailed, and insulted by Agent McGillycuddy—was up again. McGillycuddy was removed in 1886, and the old chief (with an easy going Irish agent established as his new official father) was not only recognized as head chief by the Indians at Pine Ridge, but was generally looked upon by the Sioux as the grand chief of the nation, a rank never approached by any Sioux in the memory of living men. Spotted Tail had glimpsed such a rank but had been murdered in 1881 before he could achieve the goal.

The decision of the United States Supreme Court in the case of the murder of Spotted Tail by Crow Dog had made it clear that the courts had no jurisdiction over offenses committed by Indians against Indians on Indian reservations. This decision had astonished the public and had shocked the leaders of the Indian welfare groups, who soon brought forward a scheme of their own for dealing with a situation which they regarded as intolerable. Indian courts with Indian judges were to be set up to deal with all cases of crime on the reservations. As far as the Sioux were concerned, these Indian courts had little effect when they were first established. Most of the Indians continued to settle their differences by tribal custom. At Pine Ridge, McGillycuddy turned thumbs down on the Indian court project, announcing bluntly that there were no Indians at his agency who were in any manner fitted to act as judges. This agent had made his own laws and had set up a star chamber court of his own with one of his most trusted white employees acting as judge and jury, and this police state court was the only one McGillycuddy would approve of.

Like all the other schemes that the Indian welfare leaders lumped together under the high sounding title of "progress in civilization," this plan for Indian courts was taken in a different spirit at each Sioux agency. McGillycuddy knocked it on the head and tossed it out. McLaughlin of Standing Rock welcomed it with cries of pleasure. He promptly formed an Indian court with Indian judges; but the problem of the payment of fines now came up. His Indians had little money; most of them had none; but they were rich, and, in the view of their agent, too rich, in weapons, and far too prone to use them. To remedy this situation McLaughlin promulgated a list of fines for various offenses, all fines to be paid in weapons. Both the Sioux and the Western newspaper editors were highly amused over this; particularly when the haughty Sitting Bull—chief of the former hostiles—assaulted another Indian with weapons, was brought before the Standing Rock Court and fined one Winchester and two tomahawks by the Sioux judges. By the end of the year McLaughlin's court had collected an imposing pile of assorted weapons as fines, and a majority of the Sioux at Standing Rock approved, as most of the weapons came from Sitting Bull's former hostiles, who were making themselves unpopular by parading about armed and assuming an air of immense superiority over the tamer agency Sioux.

At Cheyenne River Agency the tradition of Prussian discipline left by Captain Schwan was instilled into the new Indian court. Fines were neglected in favor of jail sentences at hard labor. Most of the Sioux whose names turned up on the crime lists were of the nonprogressive elements—men who went about armed, avoiding all forms of work, and always ready for a fight. The more progressive and peaceable Indians were pleased to see these warriors brought into court and given from ten days to three months at hard labor.

On the whole, the new Indian courts were interesting and amusing toys. Their legality was extremely doubtful. The Supreme Court itself had decreed that the government had no control over crime on Indian reservations, and the pretense that these courts were a purely Indian affair was absurd. The courts were a part of the "progress in civilization" schemes that were being planned by the leaders of the Indian welfare groups. It was pretended that the Indians alone were running the courts, when in fact the white agents were in complete control. They had the authority to set aside any decision of the Indian judges, and homicides and other really serious crimes were not even brought before the Indian courts for trial.

At Rosebud, the Brûlé Sioux were clinging tenaciously to their old customs, and their agent, James G. Wright, found it very difficult to start any bit of so-called progress, such as an Indian court. He was a cautious administrator, without a trace of the fire and originality that gave McGillycuddy and McLaughlin their rating as men of outstanding ability. Wright had set up an Indian court, as he did everything that he was ordered to do. He then left it to shift for itself, and the Sioux practically ignored it. They preferred to settle their differences by tribal custom. The Rosebud Sioux were still trying vainly to find a successor to Spotted Tail. Young Spotted Tail had hoped to be head chief after his father's murder in 1881, but he had soon found that he had little backing in the tribe. In 1882 he began to court the white officials, pretending to be a progressive. He helped the Edmunds land commission persuade many Brûlés to sign the agreement of 1882, only to lose rank among the Brûlés when they discovered that the agreement was a swindle. He put on citizen's clothing, lived most uncomfortably in his father's mansion, and rode about in a nice buggy, still courting the officials. Then he discovered that the government did not intend to recognize a new head chief at Rosebud. The young chief now decided that the thing to do was to make himself more popular among his own people. He took off his white man's garments, moved out of his father's mansion into a tipi, and began to consider a plan for making himself head chief by the good old Brûlé method of removing his principal rival, Chief White Thunder of the Loafer Band.

Crow Dog was now back at Rosebud, parading his importance. He was in his own opinion at least a big man, for had he not boldly killed Spotted Tail, spent months in prison, been tried and sentenced to hang and then released on the order of the Great Father's Great Court, which had decided that no Indian could be punished by a white court for killing an Indian on Indian land? Crow Dog and a group of his kinsmen of the Orphan Band now had log houses south of Rosebud Agency; here old Iron Shell, the last great chief of the Orphans, was buried, and near here young Spotted Tail was living. Young Spotted Tail and his close friends, considering the case of Crow Dog, decided that the way to deal with White Thunder's pretensions to leadership in the tribe was to pin a quarrel on him and then kill him. Turning Bear was the principal adviser of Spotted Tail in this matter, constantly urging him to take bold action and win back the leadership in the tribe that his father had held.

On May 29, 1884, Spotted Tail rode to the Loafer camp on Scabby Creek, six miles west of the agency, and stole White Thunder's pretty young wife. When the older chief returned home and learned of this, he was furious and acted just as the plotters had expected him to. He rode to Spotted Tail's camp and, finding him away, rounded up his pony herd and drove the animals to his own camp, where in a typical Sioux act of revenge the ponies were shot.

Spotted Tail and his friends now had the excuse they had desired. They made up a small party and went to a point near White Thunder's camp, where they set an ambush. The men in the party are generally stated to have been young Spotted Tail, "the Brûlé devil Thunder Hawk," a brainless warrior named Long Pumpkin (alias Long Gourd Rattle), and White Blanket, who is said to have been a brother of young Spotted Tail. These men hid in the bushes, and presently they saw White Thunder and his very old father coming up the trail, driving some ponies. The killers opened fire, but their first shots missed. White Thunder whirled to face them and fired, wounding Long Pumpkin severely, but the young warrior fired at the same moment, shooting White Thunder dead. The old man, White Thunder's father, was mortally wounded. The killers hastily concealed Long Pumpkin, who was too badly wounded to ride, covering him with wild vines and leaves; they then mounted and made off at top speed, fearing that men from White Thunder's camp would ride out in force and kill them. After darkness fell, Long Pumpkin crawled out of his concealment and started toward home, staggering or crawling all the way. He limped for the rest of his life.[1]

Agent Wright, on learning of this double killing, sent a man to invite young Spotted Tail and Thunder Hawk to come to his office for a talk. He listened glumly to their version of the affair, as translated by his interpreter, and then suggested that it might be well if they would submit to arrest and temporary confinement in the guardhouse at Fort Niobrara, where they would be safe from White Thunder's followers until the present excitement died down. The two worthies cheerfully agreed to this, and the Indian police escorted them to the military post in Nebraska. While they were sitting safe in the guardhouse their friends

[1] White Thunder is said to have been born in the Wazhazha camp, but after 1866 he was a chief of the Loafer Band. He had a fine reputation as a warrior and the Brûlés say that he was selected by the tribal soldiers to shoot Captain Fouts in the Horse Creek affair in 1865. He rode up to Fouts and shot him off his horse, a deed the Sioux regarded as glorious. In later years White Thunder was regarded on the reservation as a progressive. He made a good chief for his own band, but his abilities were only small, and his efforts to gain the position of head chief that old Spotted Tail had held were without result. The Brûlés of today—who know almost nothing of these early chiefs—imagine that White Thunder was killed on White Thunder Creek, a southern tributary of White River. That is a poor guess. White Thunder Creek got its name because this chief's band camped there when the Brûlés first came to the reservation in 1868. At the time of his death White Thunder's Loafer camp was on Scabby Creek, six miles west of Rosebud Agency.

at Rosebud settled for the two murders by tribal custom, paying ponies, rifles, fine blankets, and money to White Thunder's family as blood money. Spotted Tail and Thunder Hawk presently returned home, but they found that the murder of White Thunder and his old father had cost them the good opinion of the tribe. The common men among the Brûlés did not have a great deal of wisdom, but they felt that on the reservation they were living under new conditions and that it was necessary for their chiefs to have ability beyond that needed to plan and carry out a killing from ambush. Young Spotted Tail lost all followers outside his own small group of kinsmen and presently sank into obscurity.

Agent Wright was greatly distressed over this double murder. He had taken over the agency on August 15, 1883, and his coming had been hailed as the dawn of a new day of progress at Rosebud. Up to this time the Brûlés had suffered (or benefited, as many of the Indians thought) under the spineless rule of utterly incompetent agents, who let the Brûlés sit in tipi camps, eating free rations in almost complete idleness. There were no schools. The church was a tiny Protestant Episcopal affair whose slender membership was made up of white and mixed-blood families. Only a few of the mixed bloods and members of the Corn and Loafer bands were trying to farm, and the only work the fullbloods as a class had taken up was the hauling of their own supplies from Rosebud Landing to the agency, for which congenial light work the government paid them well.

Wright had been appointed largely because he was a practical farmer, and as soon as he took charge he shattered the arcadian peace that had thus far reigned at Rosebud by starting a farming boom. By coaxing, making promises, and judicious use of threats, he broke up the numerous tipi camps crowded close in around the agency. Swift Bear and his Corn Band moved to a remote farming locality, so distant that the old chief's followers were soon roaring that their ponies were being worn out, for by this curious arrangement they were trying to farm about one hundred miles from the agency and at the same time were being forced to ride to the agency every ten days or so to draw the free rations, clothing, and other supplies without which they could not live. They had become traveling farmers, spending about half their time on the trail, going to and returning from the agency. The other Brûlé bands established farming settlements to the north, northwest, and southwest of the agency, where with Agent Wright's energetic help they built log cabins. Just how much of this work the Sioux men did we will never know. Some help was certainly given by the white laborers on the agency payroll, and by long tradition women were the only manual laborers among the Sioux; yet logs were cut in the timberlands and rude cabins with earth floors and earth-covered roofs were erected in groups, dotted here and there along the creek valleys. The government provided free window sashes and free doors, as well as cookstoves and a few other needs to start the Sioux housekeeping. The great map of Rosebud, prepared by Agent Wright in 1885, shows these new cabins by hundreds. It shows Swift Bear's settlement on the north bank of the Niobrara a few miles above the junction of that river with the Missouri; the rest of the Corn Band with some of the Loafers were on Little Oak Creek, east of the agency. In this settlement old Grand Partisan of the Corn Band had his cabin, and near his house was an Episcopal mission and school and the cabin of Charlie Elliston, a man who played an important part in Sioux history in the old Fort Laramie days. Ring Thunder's settlement was on the east bank of Little White River, at the mouth of Cut Meat Creek, and one of Spotted Tail's sons and Blue Blanket had houses in this district. The White Thunder camp was near Ring Thunder's and all along Little White River and its tributaries, north, west, and southwest of the agency, the Sioux cabins stood in little groups or big settlements. Old Two Strike had his nonprogressive camp south of the agency, and far away to the northwest the big Wazhazha settlements were strung along the valley of Black Pipe Creek. On Big White River, at the northern edge of Rosebud reservation, the returned hostiles from Canada had formed settlements.[2]

[2] *Map of Rosebud Reservation* (1885), National Archives, Map No. 116, Tube 361, Index Number 20,045–16. The size of this map is three by six feet. The Orphan Band is on Cut Meat Creek; the Two Kettle Sioux who came from Cheyenne River agency in 1878 have a camp on the site of the later town of White River.

Agent Wright at once began the building of log schoolhouses in these camps or settlements, and within two years Rosebud, where no school had existed up to the time of Wright's coming, had equaled or surpassed McGillycuddy's boasted performance at Pine Ridge. The Episcopal church was also at work, building little chapels and mission schools in the new Indian settlements, and the Roman Catholics had sent a young priest to Rosebud, to look over the field. The Catholic church was planning a large boarding school in the nonprogressive districts south of the agency.

By 1885, Wright had the reputation of being one of the most efficient and valuable of the Sioux agents. He had broken up practically all of the old tipi camps close to the agency and had his Indians settled in new log cabins on valley land that could be farmed. There were fourteen log schoolhouses in the new Indian camps, and many of the Sioux were at least going through the motions of farming. To close observers it was to be noted that the plowing was done by agency white employees with agency teams, that the planting was usually accomplished by the Indian women, using the ancient Arikara and Mandan Indian hand methods of putting the seed into the ground and doing a little cultivating with hand hoes. Despite the permanent look of the log cabin settlements, these Sioux were still on the wing. They moved to new locations every one or two years, taking down and putting up their cabins as they had done their tipis in the old days. They had an excellent reason for moving. When the virgin sod was plowed up, there were no weeds the first year, but in the second and third seasons the weeds moved in—and the Sioux moved out. They took up a new location and got their father the agent to send his white men to plow up more virgin sod for them. Why break your back hoeing weeds, or your wife's back at least, when you could get a new and weedless corn patch plowed by merely asking the agent or his farmer? But Agent Wright was becoming aware of this situation and was taking, as the Sioux thought, a quite unreasonable view of it. He seemed to believe that hard work would be good for the Sioux. They did not agree, and as Wright was no McGillycuddy, ready to die on the spot or precipitate a war rather than to let any of his Sioux have their own way, the Brûlés continued to shift their locations every year or two.

Agent Wright was really trying to be patient and not expecting any miracles. He had no sympathy with the Eastern Indian welfare leaders, who seemed to know nothing about the Indians in whose lives they were determined to meddle. He reported tartly that these men should try to realize that it was not a question of providing every Sioux family with a farm of from 160 to 320 acres, but of inducing the Sioux to settle down and really work their tiny corn patches, which did not average much more than an acre per family. His very best fullbloods had fields of four acres, but the average "farm" at Rosebud was a fraction of that.

While Wright was laboring ceaselessly to get his Indians settled, to start them at work, and to establish schools, he was being subjected to annoyance from various quarters. For one thing, a young Roman Catholic priest, Father Craft, had been sent to Rosebud about 1883 to lay the groundwork for the establishment of a mission and boarding school. Father Craft seemed to think that the proper method for starting a mission was to incite the Sioux to disobey the agent. Wright reported that when Captain Pratt came to Rosebud and a council was called to discuss the sending of some additional Brûlé boys to Carlisle School, Father Craft appeared in the council and advised the Sioux not to let any of their children go, as the Roman Catholics would soon have a big school at Rosebud in which the boys could be educated, right there at home. Agent Wright would probably have passed this over, but Captain Pratt furiously denounced Father Craft, demanding that the Indian police remove him from the reservation. Wright smoothed the angry captain down and reported the incident to the Indian Office. Father Craft took full advantage of the agent's forbearance, making a real nuisance of himself. Wright now instructed his chief clerk to be on the watch, and the next time the young priest broke out to take down his words and have the document signed by witnesses. Soon after this Father Craft came to the agency when all the Indians were assembled to draw rations, and before the throng of excited Sioux he made a violent speech denouncing Agent Wright and the Washington officials.

The chief clerk took down the oration and had his copy read and witnessed. It was then sent to the Indian Office, and Inspector Tappan was sent to Rosebud (December, 1883) for a final check-up. Tappan reported that Father Craft was telling the Brûlés that the agent, the Commissioner of Indian Affairs, the Secretary of the Interior, even the Great Father himself, were small men whom the Sioux need not obey; but that he, a priest of the church, represented God, and his words must be heeded. After considering this report, the officials at the Indian Office ordered Father Craft to leave Rosebud at once. The priest sold his new house, north of the agency near the Spotted Tail mansion, to Standing Bear (a fullblood who was running the only Indian-owned store at the agency) and obeyed the order. Some Roman Catholics raised a cry of persecution, but there was no indication of anti-Catholic feeling among the officials. Indeed, they permitted Father Craft to go to Standing Rock and did not object when Agent McLaughlin, a Roman Catholic, put this young priest in charge of a government school on a government salary.[3] The Roman Catholic church sent more discreet workers to Rosebud and soon had a good mission established among the nonprogressive bands that lived south of the agency—mainly the Two Strike and Sky Bull bands. Theoretically the churches were free to establish missions and schools at all Sioux agencies; but there was a kind of gentlemen's agreement that each agency was to favor a certain denomination. The Roman Catholics were favored at Standing Rock; at Rosebud the Protestant Episcopal church was in general control; but when the Democrats came into power in 1885, they began to favor the Roman Catholics at all agencies. At Rosebud the Episcopalians thought it only fair that the teachers in the new government day schools should be mainly Protestants. Whenever a new teacher was to be appointed, the Episcopalians reminded the Democratic officials in Washington of the situation at Rosebud, and in almost every instance their representations were ignored and a Roman Catholic teacher was sent out.

At Pine Ridge, Red Cloud was demanding a Roman Catholic

[3] See *Report of the Board of Indian Commissioners* (1890), 179–80; also the Wright Reports (1883–84).

mission mainly, it would seem, to annoy McGillycuddy, who was working in close co-operation with the Protestant Episcopal church. That church had been given a large fund for mission work among the Sioux by the will of the late Mrs. John Jacob Astor, and with this money handsome little chapels had been built in some of the Sioux camps at Pine Ridge. McGillycuddy was all for the Episcopalians, and when a Roman Catholic priest came to Pine Ridge to see what the prospects were for a mission, the agent warned him off, stating that if he persisted in attempting to start Roman Catholic work, he would be put off the reservation by the Indian police. In 1886 McGillycuddy was himself removed, and the Roman Catholics came in at once and started a mission for the Red Cloud and other nonprogressive groups.

In these years the big matter at all the agencies was farming. McGillycuddy told the Washington officials bluntly that neither the soil nor the climate at Pine Ridge were at all favorable to Indian farming and that the government plans were folly. He did give some encouragement to his Sioux to plant small patches of corn and vegetables; but in his view stock raising was the only hope for making the Pine Ridge people ultimately self-supporting. At Cheyenne River Agency a reverse opinion was held. Captain Schwan had forced his Sioux to farm from 1878 until he left in 1880. Drought usually destroyed the little crops, but the agent believed that it was good discipline for the Sioux to work hard, crops or no crops. His Indians violently objected to that view of it; but when Schwan left they were still forced to keep up the pretense of farming.

In 1883 they had a new agent, William A. Swan (no relation to Captain Schwan) who was inclined to let up a bit on the Sioux. That summer by some miracle large herds of buffalo drifted into the western borders of the Sioux reservation, and the Cheyenne River Indians were wild with excitement at the opportunity to go on what might be their last buffalo hunt. Agent Swan wished to be kind; but he had orders from Washington not to permit anything to interfere with the sacred farm program. He finally consented to let a chosen few from each of his Indian camps go on the hunt while the majority stayed at home and farmed. He was

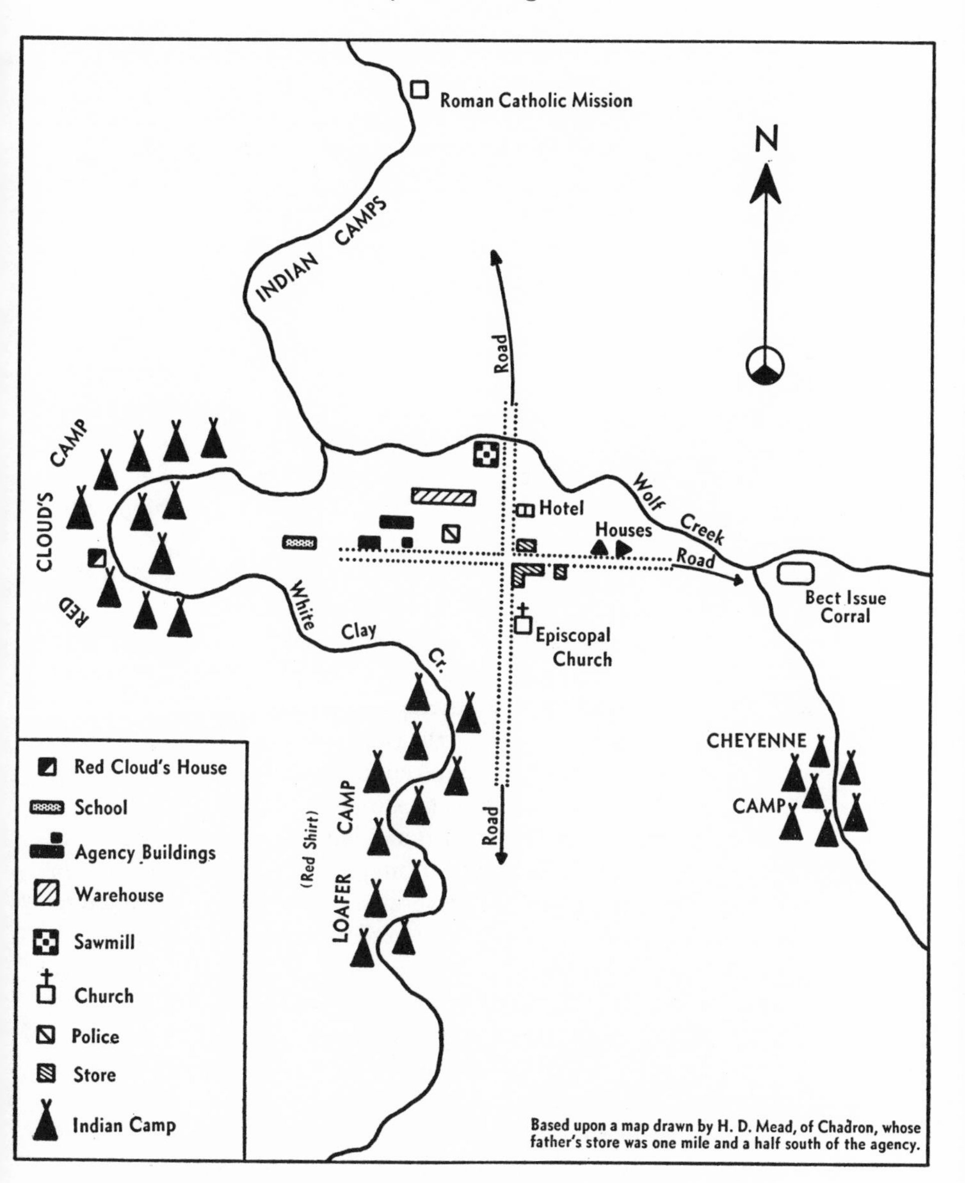

Pine Ridge Agency and Vicinity

unwise enough to inform the Indian Office of his plan, with the result that he received a telegram reprimanding him and ordering him to bring the hunters back at once to their farm work. Swan chased his Sioux up Cheyenne River, caught them within sight of the buffalo herds, and forced them by threats to turn sullenly homeward. These Sioux could have supported themselves for weeks by hunting, also bringing home great quantities of dried buffalo meat and the hides, which they could have exchanged at the traders' stores for goods they needed. They were deprived of all this and of the great pleasure of a last buffalo hunt, and the result was that in autumn Agent Swan reported: population, 3,214; acres planted by Indians, 150; crops, "ruined by floods." The farming program had been carried out, and this was the result. If it was not drought at Cheyenne River, it was floods, for all the best farmland was in the stream valleys and subject to overflow in high water. Of course there were good years; but even then the Sioux grew only a fraction of the food they ate. The rest came from government rations.

When the troops disarmed and dismounted the Cheyenne River Sioux in 1876, the Indian ponies were driven to the white settlements in Minnesota and southern Dakota and sold at very low prices. The military employed the funds obtained from the sale of the ponies to purchase six hundred cows and eight bulls which were turned over to the Sioux, to start them at stock raising. The Indians preferred herding to farming, but—with little thought for the future—many of them when hungry killed their stock cattle and ate them. Strict rules and severe punishments were invoked to stop such practices, and the little herds at Cheyenne River grew. By 1890 the Indians there were producing one-fourth of the beef required to feed themselves. It was a curious form of self-support. They sold their beef to the government; the government then issued the beef as free rations, adding other free rations to make up the full amount required.

But even stock raising was difficult in western Dakota. By 1886 the white squawmen and mixed bloods, who had the best herds on the reservation, had found that the open range system would not work. Their cattle had to be sheltered and fed in winter. In western Dakota the open-range advocates were in the majority, but the year 1887 put an end to that. The drought was very severe. It reached all the way down into Texas, and as the season wore on, southern cattlemen began to drive their hungry herds northward in quest of better grass. In Dakota the drought was so bad that even the grass on the range was very poor; but by the end of autumn the range was crowded by immigrant herds brought in from the south. Then came the winter of big snows and intense cold, and when spring came the range was covered with dead herds. Most of the big cattlemen were ruined, and the open range cattle industry never recovered from the blow.[4]

By 1886 it was apparent to men who were not blinded by the faith of crusaders that attempts to hustle the Sioux suddenly into self-support from the soil was bound to fail. Ultimately—not in a few years, but in generations—these Indians might achieve a rating as subsistence farmers; but as a people they would never have any large marketable farm surpluses. They might do better with stock; but even there it was a question of a long period of years. Part of the Sioux were certainly getting used to the idea of a little manual labor, and men who had known them in 1868 when they were still wild buffalo hunters and warriors might be justified in describing their progress on the reservation as good. In 1886 the agents' reports showed that at four agencies about one-half of the families were cultivating gardens (not farms), and McLaughlin of Standing Rock claimed that all his families were trying to farm. The figures showed that the family farms averaged from two acres up to almost four. This may seem absurdly little; but there are the records of tribes that practiced growing crops for centuries before the whites came into the land, records that show how well the Sioux were doing after only twenty years of effort. The Pawnees of Nebraska had grown crops since prehistoric times, but in the 1830's J. B. Dunbar estimated the average Pawnee corn fields as from one to three acres per family. The Arikara tribe (cousins of the Pawnees) had been growing crops on the Missouri River in Dakota as early perhaps as the year 1600; yet, after being pushed and hustled by government agents for

[4] *South Dakota Historical Collections*, XIV, 154.

twenty years, the Arikaras in 1878 had an average of less than three acres per family in crops. This tribe of ancient cultivators had formerly obtained most of their food by hunting. With hunting ended, they were in 1878 no farther advanced than the Sioux, obtaining 15 per cent of their food from their little fields and 85 per cent in free government rations.[5] This fact, that many tribes that had been growing crops before Columbus came had not bettered the acreage developed by the Sioux in twenty years' effort, caused some thoughtful men to wonder if these Indian tribes had not reached a kind of saturation point in the matter of farming, beyond which they would never go. They just did not seem competent to develop farming on the white man's standard of big crops to be marketed.

It is a curious fact that none of the Sioux agents of this period (whose reports were largely taken up with matters of rationing, farming and self-support) ever had a word to say concerning the diet of their Indians. One agency physician, at Cheyenne River, did send in a blistering report on this subject. He stated that, after all the expense and labor that had been put into the farming program, the Sioux still did not have enough vegetable foods to make a dent in their steady diet of meat and pan bread. The main articles on the list of free rations were low-grade beef, low-grade salt or pickled pork, and a very low-grade flour, specially milled as Indian flour and of a quality that no white family would use. The cheapest grade of green coffee beans were added, and rough brown sugar. This physician reported that he had never known an Indian woman to roast, bake, or fry meat; everything went into the huge tin kettle that was always simmering: meat and any vegetables that could be obtained. The Sioux preferred shelled corn, as they could boil the corn with their meat or make it into hominy. A great deal of the flour ration was wasted; the rest was made into tough pan bread, baked in a skillet. This diet, which was enough to kill unaccustomed white men, was the regular portion of all the Sioux, from babies to great grandmothers, year in and year out. It apparently did not affect the Sioux, their main

[5] George F. Will and George E. Hyde, *Corn Among the Indians of the Upper Missouri* (Cedar Rapids, Iowa, 1917), 103.

complaint being lack of quantity. Their habits were still those of hunters, and when ration day came most of the families started feasting and inviting friends to come and eat with them. They consumed their rations, intended to last them for a week or ten days, in three or four days, and then squeezed through on any food they could find until next ration day. They had lived like this in the old hunting days, feasting after a big buffalo kill, and then going hungry until they had the good fortune to find new herds to hunt.

When the Democrats at the Indian Office removed McGillycuddy of Pine Ridge in 1886, their explanation was that he had refused to obey the orders received from Washington; but when they replaced him by appointing a Democrat, their true object was disclosed. Other Republican agents were put on their guard, while in the East the Indian welfare groups lifted their voices in loud protest and prepared to fight in defense of these agents. In particular they were bent on defending Agent Wright at Rosebud; but Wright was not a fighting man like McGillycuddy, and he preferred to do nothing further except to walk with great circumspection, hoping that if all was quiet at Rosebud the Democrats might overlook him.

At his agency there was a big warehouse filled with furniture, stoves, crockery, and other household articles, including even sewing machines, which Wright was using as bait to induce his Sioux to work and to progress in civilization. If an Indian built a log cabin, he was given a cookstove and some plain furniture; if he built a two-room cabin, he received in addition to those articles a heating stove, extra furniture, a table lamp, and certain other articles; and if he and his wife were both good workers, they were given a sewing machine. This system of rewards worked well and was Wright's pride; but now it was to prove the ruin of his wish to keep his agency very quiet, for over on Little White River six miles west of the agency was Chief Wooden Knife's tipi camp, a purely nonprogressive group whose men were content to eat free rations and wear free blankets in idleness. But the chief's wife was a progressive. She longed for a sewing machine and she kept talking about it until in sheer self-defense the chief went to Agent Wright and asked for one. Wright told him that if he left his tipi camp, put on white men's clothes, built a two-room cabin, started farming, and performed a large number of other laborious acts, his wife would be eligible for a machine, but until he was ready to qualify fully it was no use for him to hang about talking to a man who was very busy.

Wooden Knife was a warrior of some renown and had a good opinion of himself. He did not like Agent Wright's manners and did not like the arbitrary standards of good conduct that this white man had set up. They were not good old Sioux standards at all, and when he went home and talked with the other men of his camp he found that they agreed with his views. They also agreed when he told them that something must be done, for every man of them had a hard-talking wife who wanted a sewing machine. They held councils and unanimously decided that all the fine articles piled up in the warehouse were a payment to all the Sioux for lands given up and that no white man had the right to make rules and compel any man to pay for his own goods a second time by building a cabin and hoeing corn.

On a day when all the Sioux were assembled at the agency, the Wooden Knife men rode in, picked up Louis Roubideaux to interpret for them, and crowded into the office, where they cornered Agent Wright; but he resisted all their demands for sewing machines and other luxury articles, and after a furious debate punctuated with Sioux yells and threats the Wooden Knife band rushed outside, to tell the assembled Indians loudly of the bad treatment they were receiving. Crow Dog presently got hold of them and took them to his cabin for a council. This scheming Brûlé was now posing as a kind of reservation lawyer, his experiences with legal procedure having been extensive, for after his murder of Spotted Tail he had spent many months in prison, had been tried and sentenced to hang, and in the end had been set free on orders from the Supreme Court. Thus, when any trouble arose at the agency, he had wise counsel to offer the Sioux. He now advised the Wooden Knife men to stand fast in their quarrel with the agent, and he assured them that the Great Father had said to the

Sioux chiefs that if an agent did not act properly, the Sioux were to pull him out of his office.

On May 1, when the Sioux assembled at the agency, the Wooden Knife warriors rode in to renew the squabble with their father the agent; but Wright had heard rumors of the meeting at Crow Dog's place and he now had all his Indian police on duty, under strict orders to prevent any disturbance. The moment the Wooden Knife faction rode in, the police asserted themselves; the trouble makers withdrew and all seemed well. But immediately thereafter Agent Wright's real education in Sioux ways was begun.

The Wooden Knife warriors rode over the hills west of the agency and vanished, but presently from behind those hills came the faint, far-off sound of chanting. Then over the hills came the Wooden Knife band, mounted and armed, singing "brave songs" as they advanced. The Sioux who thronged the agency ran this way and that, shouting; the Indian police drew together and attempted to stop the advance of the Wooden Knife men, but they were brushed aside. Followed by a mass of yelling Sioux, Wooden Knife made a dash for the big council hall which was connected by an inner door with the agent's office. In an instant the council hall was jammed with howling Indians: Wooden Knife and his followers, some of the police, warriors from bands that favored the agent and from bands that supported Wooden Knife. They were pressed together like sardines in a box, all yelling into one another's faces and struggling to get an arm free, to reach for a knife or lift a rifle.

Some of the police fought their way to the door that led into the agent's office and there made a stand, trying to force back the mass of angry Sioux. The door suddenly opened and Wright stood in it, gazing at this shocking scene of riot. He held up a hand, demanding silence, but it was long before the Sioux stopped fighting each other and it was possible for the agent to be heard. He spoke and his interpreter translated his words to the Sioux. He said that if anyone wished to speak to him as a friend, it was well; but to men who were not friends he had nothing to say. Wooden Knife thumped a war club on the floor and shouted that the Great

Father had once told the Sioux that if an agent did not treat them well, they should throw him out, and now he and his men were going to do just that. Wright demanded a paper—a document showing that the Great Father had ever said such a thing. Wooden Knife thumped the floor and shouted that he had no paper. The agent angrily declined to listen further. He turned his back—a bad error. An angry Sioux grabbed his right arm, a worried policeman his left, and they began to pull and haul, the unfortunate agent bawling protests as he was whirled round and round, the center of a little solar system whose rim was composed of clusters of battling Sioux, many of whom were yelling fiercely for someone else to kill the agent.

A little group of policemen fought their way through and pushed Agent Wright into his office: a good move, but one that was hard to stop after it had once started, for a solid mass of Sioux trod on the heels of the police, and Agent Wright was propelled rapidly across his office, through a second door, and out into a yard. Here the Sioux and the policemen had more space to maneuver in, and they fell to fighting again. Wright was hustled off to safety by some of his allies, but the mighty squabble between his friends and those of Wooden Knife roared on and on. It was the arrival of the chiefs with the tribal soldiers that finally ended the fighting. They were men for whose authority the Sioux had a proper respect, and the most enraged of the combatants broke off the fighting when their chiefs and soldiers ordered them to do so. But even with the fighting ended, the confusion at the agency was prodigious. The Sioux gathered in angry groups and quarreled violently. They followed the harassed Indian police about jostling them and making threats. Darkness had fallen on the scene of wild confusion; some of the Indians had cornered the warehouse clerk and taken his keys from him, and a mob of Sioux now went to the big darkened warehouse, opened its doors, and began to take out treasures: sewing machines, table lamps, handsome crockery, and even bedsteads. They might have emptied the warehouse if some of the chiefs had not come with their soldiers and driven the mob off. To the Sioux this affair was always known as the "night issue"—the only time in all history when articles were issued at the warehouse after dark. They called it an issue of goods, when in simple truth it was looting.

The curious part of all this was that no one was seriously injured. All the hundreds of angry Sioux had rifles, Colt revolvers, knives, or war clubs, but they also had the queer Sioux ability to lose their tempers completely and fight furiously while still holding firmly to the fundamental fact that they must not kill a fellow tribesman.

This affair was never settled properly—not at least according to McGillycuddy's idea of a settlement. He would have had every one of the offenders in jail within the hour or would have died right there; but Wright, according to his own mild version, demanded an apology which the principal culprits gave him. The Sioux version was to the effect that the chiefs, after stopping the fight, held a council with the agent, who finally agreed to settle the matter by giving Wooden Knife and all the men in his camp a share in the goods from the warehouse which they had already taken without his permission. Agent Wright tried to gloss over this uproarious performance as a small incident. It was, however, one that rang from end to end of the great Sioux reservation. Not since the day in 1873 when the Red Cloud people had dragged their agent, Dr. J. J. Saville, off his horse and tried him before a tribal council had the Sioux laid violent hands on the sacred person of one of their official fathers. They had not only laid violent hands on Wright, they had laughed about it when their anger had cooled, and for years later the Sioux fell into helpless mirth whenever they were reminded of that scene, with Wright being waltzed around, an enraged Sioux yanking at one of his arms and an anxious policeman clinging desperately to the other.[6]

Agent Wright did manage to keep this affair from being widely known outside the reservation, and the Democrats permitted him to finish his term of six years. Then they appointed a Democrat as agent at Rosebud.

[6] Old Sore Eyes said that it was Eagle Pipe who pulled Agent Wright out of his office, by which he evidently meant that Eagle Pipe had one of Wright's arms while the Indian policeman had the other. The agent's idea that this was a minor event is refuted by the fact that all the Sioux winter-count records set down the pulling of the agent out of his office as the big event of 1886.

7

Never Call Retreat

FROM THAT day in the spring of 1885 when Secretary of the Interior Lamar had told them to their faces that he had no faith in their proposals for rushing the Indians suddenly forward into the ownership of individual farms and into full citizenship, the leaders of the Eastern Indian welfare groups had lost faith in President Cleveland's Democratic administration. But they did not give up. They belonged to the class of embattled idealists who never call retreat but stand and fight to the last. Having lost hope that the Democratic officials would do their bidding, they took the fight to Congress; but that body was strangely backward in taking the action that the leading Indian Friends regarded as of most immediate and vital importance. Their demands for great increases in appropriations for Indian education met with no response. The land in severalty bill (providing for a farm for each Indian family, for the sale of surplus Indian lands, and for pressing the Indians along swiftly into full citizenship), failed to receive any important support. Senator Henry L. Dawes of Massachusetts had written this bill and was leading the fight in the Senate for its enactment; but as the struggle for its passage dragged on and on, he became hag-ridden with doubts concerning the wisdom of the measure. This was a very powerful piece of legislation, of the type that doctrinaires usually concoct, picturing themselves in control and desiring to have all possible powers put into their hands, to be employed in a worthy cause. This was all very well for firm believers of the Indian Friend groups, but Dawes saw visions of what might happen if this law with all its powers should come into the hands of bad men who would use it to destroy instead of to aid the Indians. Coercion had been written into every part of this law on the peremptory demands of the Indian welfare leaders; when Dawes and other men of sense protested that such a measure could not be defended, Dr. Lyman Abbott—a leading Indian crusader—stated excitedly "that any means could be defended that would help attain the goal" of forcing Indians to progress. Dawes tried to talk reason to Dr. Abbott but failed to impress him. When Herbert Welsh tried his hand, he reported to Dawes that he had failed to alter "the crude and radical views expressed by Dr. Abbott and Dr. Rhoads."[1]

Even under such conditions, Senator Dawes did not see his way to draw back. He pressed the land in severalty act to a final vote, and it was passed on February 8, 1887. The Indian welfare crusaders greeted its passage with cries of joy. They fondly termed it the Indian emancipation act, a truly strange label for a law which was purely coercive, which gave the government the power to violate Indian treaties, to take the Indians' land without their consent, and to turn over all Indians, tied hand and foot, to agents to be driven like slaves.

Senator Dawes had also introduced a new Sioux land-purchase bill, under the terms of which the whites of Dakota would get possession of the same amount of Sioux land that they had attempted to take under the Newton Edmunds land agreement of 1882. Dawes honestly intended this land bill to be an aid to the Sioux. He had no illusion that these Indians would suddenly go to work and become successful farmers; but he was convinced that, much as they might oppose the sale of their lands, the 25,000 Sioux could not hold out much longer against the pressure exerted by the 500,000 whites in Dakota, who were trying to get the Indian lands. The senator's view was that unless the Sioux gave up part of their lands to satisfy the whites, they might lose the whole reservation. The leaders of the Indian welfare groups eagerly supported this bill, their object being to get rid of all Sioux land except for a farm for each Indian family. Thus with the Sioux pinned down on farms the welfare leaders confidently

[1] Priest, *Uncle Sam's Stepchildren*, 245.

expected that they could put pressure on the Sioux and force them to go to work. These idealists imagined that they knew what was best for the Indians, and they were determined that the Sioux should do their bidding. This bill was intended to please all parties concerned. It gave the Dakota whites the Sioux land they desired, it protected the rights of the Christian missions in their holdings of church and school lands on the reservation, and it gave the Indian Friend groups full scope for meddling in the lives of the Indians and attempting to force them to progress. The bill had the hearty support of everyone—except the Sioux Indians, the Democratic administration, and the public. Pushed vigorously by those who would benefit from its passage, the bill made no progress. Secretary of the Interior Lamar was opposed to this attempt to hustle the Sioux forward, and the Democrats in Congress accepted his views and would not support the bill. Lamar regarded with deep suspicion the effort of Dawes and the Indian welfare leaders to label half the Sioux reservation surplus land and sell it off. He had little faith that the Sioux would ever farm on a commercial scale in the drought-stricken country of western Dakota; and if these Indians ultimately had to try to make a living by growing cattle, they would need the lands that this bill proposed to sell for a song.

It was when the Cleveland administration got into difficulties and needed Republican support in Congress that the Sioux land bill began to move. Lamar resigned. There was a new secretary of the interior; the Tennessee Democrats lost control of the Indian Office, and Cleveland appointed a new commissioner of Indian affairs who was on friendly terms with the leaders of the Indian welfare societies. A new Congress was elected, the Sioux bill was passed, and the Indian welfare leaders took heart again. This was the best opportunity they had had since Cleveland's election in 1884 to meddle in Sioux affairs, and they were ready with half a dozen new schemes for making the Sioux over along lines of their own devisement.

The Dawes bill, passed April 30, 1888, was entitled *An act to divide a portion of the reservation of the Sioux Nation in Dakota into separate reservations and to secure the relinquishment of the*

Indian title to the remainder. It provided, as the Edmunds agreement of 1882 had, that the great Sioux reservation held in common by these Indians should be divided into six small reservations, one for each Sioux tribe or group; the surplus lands left over after the six reservations were set aside were to be ceded to the government and opened to settlement. As in the Edmunds agreement, some nine million acres were to go to the whites; but, unlike the Edmunds plan, these lands were not to be given free to settlers but were to be sold at prices from 50 cents an acre for the poorest land up to $1.50 for townsite land. The Indians were not to receive this money, but it was to be set aside for their benefit, the officials in Washington to be the sole judges as to what would benefit the Sioux. As additional inducements for signing away their lands the Indians were promised an extension for thirty years of the educational advantages provided under the treaty of 1868; they were to receive some 26,000 head of stock cattle, and for each family that settled on a farm two cows, a yoke of oxen, farm tools, twenty dollars in cash, and seed for five acres for two years. Eighteen thousand dollars were voted to pay the expenses of a commission which was to take this land agreement to the Sioux for approval.

Dawes was proud of this land act, which he regarded as a model agreement. He believed the terms offered to the Sioux were so liberal that the Indians would sign at once, willingly and gladly—which goes to show how little even United States senators who specialized in Indian matters really knew about the Sioux. Ever since 1882 these Indians had been united in bitter opposition to the sale of any of their lands on any terms whatever. All these years they had been watching and had kept the squawmen reading newspapers, to find what the Great Father's council was planning to do. They had feared that another land bill would be introduced, and when they heard that the Dawes bill had appeared before Congress, some grew excited and angry and some others prayed daily that the bill would not pass. Now that it had passed, they girded themselves for battle, and first they began the work of close-herding the weak brothers who might be talked by white men into signing away their lands. After all, there were a few things in the Dawes act that certain of the Sioux would wish to have. There was the twenty dollars spending money that the head of each family would get; there was a wagon for each family, and the Sioux dearly loved a nice farm wagon, to travel about in and visit relatives. There were a few Sioux fullbloods and many mixed bloods who were really trying to farm and would appreciate the aid promised in the bill; but from the point of view of the average Sioux of the day, the Dawes bill contained almost nothing desirable—an amazing fact when one recalls that Dawes had spent several weeks in 1883 on the Sioux reservation, cross-examining the Indians as to their views and wishes. He had now drawn his bill under that old misapprehension of the Indian welfare leaders, who imagined that Indians were exactly like whites. Any poor white family would be overjoyed at the gift of a free farm, with work oxen, tools, a wagon, and free seed for two years; so, reasoned Dawes, the Sioux would snatch at such an offer. But the Sioux were not like that at all. The great majority of these Indians had no interest in trying to farm. They believed that free rations and clothing were due to them under the treaty, in payment for lands already taken by the whites, and their attitude was: let the white men pay for the lands they have taken in the past before they come to us asking for more of our land. The educational advantages proffered in this bill, a feature that Dawes believed would prove most attractive to the Indians, only annoyed and alarmed most of the Sioux. Why pretend that these Indians were eager to see their children educated when it was a notorious fact on the Sioux reservation that only by making constant use of the Indian police to hound and frighten the parents could the little day schools be kept running? If the policing had been stopped, the Indian schools would have been empty.

Senator Dawes would have been horrified if he had known of the spirit in which the Sioux received the news that his land bill had passed and had been signed by President Cleveland. When the chiefs learned of this dire happening, they sent off runners to spread the alarm, and soon from every agency parties of chiefs and headmen were riding toward Rosebud, to hold councils and to decide on a plan for defeating this new attempt of the whites to take their lands. George H. Spencer, who had succeeded

Wright as agent at Rosebud, reported that all through the spring delegations of Sioux kept arriving at his agency for councils and that his Indians became so excited over the coming of all the great chiefs and famous men of their nation that they dropped their half-hearted efforts to farm and did nothing but stand around and talk. The angry agent sent his Indian police to break up several of these councils; but the Sioux simply slipped away in small groups and held new councils in more distant camps. At all of the agencies the excitement was great and farm work was neglected while the Indians held councils. By early summer it was known to all of the Sioux agents that their Indians were determined to defeat the Dawes agreement; yet in the East no one seemed to doubt that the Sioux would sign willingly. The officials and the leaders of the Indian welfare groups had agreed on the ideal person to take this document to the Sioux and induce them to sign it. They had chosen Captain R. H. Pratt of the Carlisle Indian School!

They knew Pratt and admired him, hence, by their method of thinking, the Sioux all knew Pratt and admired him. In truth, only a few of the Sioux knew who Captain Pratt was, and most of the Sioux who had seen Pratt spoke of him as the saucy white man who now and again came out to the agency, got the agent to call a council, and then attempted to coax or bully the Sioux parents into letting him take their boys to his big school, some place in the white men's land, far away. Some of the boys came back years later, turned into imitation whites, and most of them were unhappy. Some died off there in the white men's land and were never seen again. That was what the few Sioux who knew Captain Pratt thought of him. The men in the East who were in close touch with Pratt should have realized that he was not at all the type of man for carrying on delicate negotiations. He knew only one method of negotiation—that which he used in dealing with recalcitrant Indian boys at Carlisle. The moment anyone opposed his will be grew angry. He was a narrow doctrinaire, absolutely wedded to his own peculiar views as to how Indians should be dealt with.[2] In

[2] A short time before he was appointed as chairman of this Sioux land commission, Captain Pratt had delighted the more radical of the Indian welfare leaders by declaring violently that Indian treaties should be destroyed and the

1882 the Edmunds commission had attempted to bully the Sioux, and the leaders of the Indian welfare groups had been horrified that such men should be sent by the government to negotiate a land agreement. Now the crusaders were delighted that Captain Pratt was to be chairman of the group which was to persuade the Sioux to sign away their lands. They were also pleased at the appointment of the Reverend William J. Cleveland as the second member of this commission. Cleveland had been a missionary at Rosebud for many years. He knew the Sioux intimately; but he was what these Indians termed a small man, not important, and he was of such a mild disposition that it was a foregone conclusion that he would at once submit to Captain Pratt's domineering temperament. His fluent knowledge of the Sioux language would prove very useful to Captain Pratt, who would employ him as a kind of official mouthpiece in speaking to the Indians. That would be the extent of Cleveland's services. The third member of this commission was John V. Wright, a professional treaty maker who was constantly employed by the Indian Office in the negotiation of land agreements with various tribes. He was a skilled routine worker, nothing more.

In the summer of 1888 this land commission took the field, going first to the northern agency, Standing Rock. Here Agent McLaughlin had received instructions from the Indian Office to do all in his power to aid the commission, but he could do nothing. He knew that his Indians were united in opposition to the sale of any land and that the only way to break down that opposition was to resort to the methods of the Edmunds commission of 1882 and threaten the Sioux until in a sudden panic they broke and signed the agreement. But both his orders and his own common sense warned him not to resort to any such methods. He called a council of his Indians, and the Reverend Mr. Cleveland of the commission read the land agreement in the Sioux tongue and explained how very favorable the terms offered to the Indians were. Chief Grass, the most notable progressive Indian at Standing Rock, then got up to speak for the tribe; but to the amazement of

Indians forced to submit to plans which the welfare groups were making to drive the tribes to progress rapidly. Priest, *Uncle Sam's Stepchildren*, 244.

Courtesy Nebraska State Historical Society, Lincoln.

Sam-Kills-Two (also known as Beads) working on his winter count

Major General Nelson A. Miles. Reprinted, by permission, from John F. Finerty, War-Path and Bivouac, or, The Conquest of the Sioux *(Norman, 1961), p.* 227.

the commission he said nothing about the land agreement they had brought, launching instead into a long discourse on the treaty of 1868, setting forth point by point the many promises in that treaty and stating how the government had failed to give the Sioux benefits it had promised them. He ended by telling the land commission to return to Washington at once and inform the Great Father that after the government had executed all the promises made under this old treaty, the Sioux might be willing to talk about the new land agreement.

This speech enraged Captain Pratt. Here was open opposition, which he would not tolerate. He started bullying the Indians.[3] He said that he would force them to vote, and to vote right. Here were white tickets, he said, printed in red (the Sioux color symbolizing life and happiness), indicating a yes vote; here were white tickets printed in black (a bad color!) for any misguided Sioux who dared to vote no. The Sioux must choose. But they would not! Who was this man who talked angry and tried to treat them like small boys? Few of them knew him; those who did knew that his business was to rule Indian boys, and now he was trying to deal with their chiefs as he did with boys! The Sioux chiefs listened to him in silence, reserved and dignified. Pratt stormed on and on; some of the Indians interjected sharp remarks; but the council broke up with nothing accomplished. Pratt summoned further councils and the same performance was gone through with. A Minneapolis reporter who was covering the councils annoyed Pratt by sending to his paper reports which the captain regarded as biased. Knowing no restraint, Pratt ordered Agent McLaughlin to remove this man from the reservation. The Indian police were called in and the reporter was taken across to the east bank of the Missouri, outside the reservation; there he camped, visited by whites and Indians from Standing Rock every day and continuing to write reports, even more biased than before. By this time the border was in a flame, angry newspaper editors accusing Pratt of being a bungler and worse. The Eastern papers took up the cry. But at Standing Rock the obstinate lord of Carlisle School butted his head against the solid wall of Sioux opposition

[3] *The New York Tribune* termed Pratt's methods at this council "bulldozing."

and would listen to no advice. This went on for two mortal weeks. Then Chief Grass arose in council and said that the Sioux had had enough and were going home to their neglected farms. The Sioux took down their tipis and left the agency.

The Pratt commission had failed. Everyone knew it, even the wrathful Captain himself. The commission did not dare to go to another of the big agencies to face the certain repetition of the Standing Rock fiasco. Word had come from all those agencies that the Sioux had counciled among themselves and stood in absolute opposition to the land agreement; but from the little agencies—Crow Creek and Lower Brûlé—reports stated that the Indians were at any rate open to persuasion, so the commission left Standing Rock and went down the river to Crow Creek.

Here they discovered that their information was incorrect. The true situation was that the Crow Creek Indians, who had voluntarily given up their chiefs some years back to conform to the government policy of dealing with all Indians as individuals, had now recalled their chiefs and the whole tribe was standing united in opposition to signing away any Sioux land. All that the chiefs had to say to the Pratt commission was that they wished to have word sent to the Great Father that the Sioux chiefs were anxious to go to Washington, to speak to him in person about their land troubles. Captain Pratt set his jaw and went to work on these Indians in the best Newton Edmunds and Judge Shannon style. He forced a vote; but the count showed 120 men voting "yes" and 282 "no," and it required a three-fourths majority to approve the agreement. Crossing the Missouri to the Lower Brûlé Agency, another vote was forced and a victory won, the Sioux voting 244 "yes" and 62 "no"; but there was a mystery here. It became clear later on that the Lower Brûlés were as much opposed to the sale of any land as the Sioux at the other agencies were. The Pratt commission had just met defeat at Crow Creek within sight of Lower Brûlé, and men suspected that the commission had privately offered special inducements to the Lower Brûlé Sioux to bring them around. From what next happened one might suppose that the commission had promised to permit a general council and to give Lower Brûlé the honor of being the place where the

Sioux chiefs should assemble. For now the Indian Office officials suddenly reversed their stand, that under no conditions should the Sioux be permitted to meet and decide on a common course of action in this land deal, and orders were hastily dispatched to all the Sioux agents to assemble delegations of leading Indians and to take them at once to Lower Brûlé.

Captain Pratt sat at that agency, watching morosely as one band of Sioux after another marched in and put up their tipis until the whole plain was whitened with Sioux camps. A big awning was put up in front of the agency buildings to form a shade, and when the council assembled, the chiefs sat under the awning while a mass of excited Sioux crowded in on every side, as closely as the Indian police would permit them to stand. Pratt opened the council, speaking sharply and, as many thought, threateningly. He said that the chiefs had been brought here to take steps to end the bad conditions under which the Sioux had been living and to bring the tribe quickly up level with the whites. As for the Sioux request that their chiefs be permitted to go to Washington, the Great Father's reply was that he did not wish to see the chiefs until this land matter had been settled to his satisfaction. At present he regarded the chiefs as bad leaders who had failed to induce their people to keep their children in school and to go to farming to support themselves. The Great Father's patience with the Sioux was nearly exhausted.

The Sioux were not particularly good at understanding white men's ideas, but they saw Pratt's meaning. The white men were in trouble and were trying to blame the chiefs. For many years the officials had told the Sioux that the chiefs were of no more importance than common Indians and they had denied the chiefs their tribal and treaty right to leadership. Now the officials' fingers were burnt, and it was all the fault of the chiefs.

Judge J. V. Wright, another member of the land commission, followed Pratt with a heavy and dull speech, talking down to the chiefs as if they were small boys. The commission then formally demanded that the chiefs sign the land agreement and pledge themselves to urge all the Sioux to sign it. They refused, and that ended the matter.

Captain Pratt wrote a brief but vitriolic report in which he recommended that treaty stipulations should be brushed aside and the Sioux lands taken without further negotiations. The *Word-Carrier*, a little Sioux mission paper, stated that the Pratt report indicated that (having failed in his task) Captain Pratt was urging the officials to put pressure on the Sioux and, if that failed, to break the treaty of 1868 and take the lands by force. Pratt was in a fine temper, striking blindly at the Sioux. Going miles out of his way as a land commissioner, he recommended that bacon should be substituted for fresh beef in the Sioux ration, both to save public money and to prevent the cruel butchering method of the Indians (they turned the cattle loose and hunted them on horseback, like buffalo, shooting them with rifles). The *Word-Carrier* quoted a former Sioux agent who stated that the ill effects of salt pork diet on Indians was notorious, and the government instead of bacon had better issue arsenic and get the poisoning process finished with decent expedition. This little mission publication spoke with much authority on matters concerning the Sioux, and it now passed judgment on Captain Pratt as one whose notions for dealing with the Sioux were "a few of them good, some criminal, most of them impracticable."[4]

During the final weeks while the Pratt commission was concluding its futile efforts to force the land agreement on the Sioux a change was taking place in Washington. Men who had their minds made up that the Sioux must for their own good be compelled to give up half of their huge land-holdings were exerting fresh pressure on the Cleveland administration. Senator Dawes was one of these men. He feared, as he stated, that if the Sioux did not give in and sell part of their lands now, they might lose all the lands before long. Dakota had won her fight for statehood and was entering the Union as two new states. The Dakotans were more anxious than ever to obtain the Sioux lands; they wanted them quickly, and they were completely out of patience with official bunglers. Captain Pratt, indeed! Was the

[4] *The Word-Carrier*, December, 1888. The Pratt report is in 50 Cong., 2 sess., *Sen. Exec. Doc. No. 17*. See also Elaine Eastman's book, *Pratt: The Red Man's Moses* (Norman, 1935).

man fit for anything more than hectoring Indian boys at his school? Both in the Dakota newspapers and by word of mouth in Washington the Dakota men expressed their views, shocking the officials by certain frank statements to the effect that they wanted real men sent out to deal with these stubborn Indians—men who would make the Sioux dizzy and end this nonsensical performance. Some of the Indian welfare leaders were critical of Dakotan ethics, but whether they approved of the Dakota men or not, the members of the Eastern humanitarian groups were now, in effect, their allies. They were united in an effort to take the Sioux lands, the Dakotans frankly for their own benefit, the humanitarians for the alleged good of the Sioux. By this time the Sioux had no friends left among the whites except Dr. Bland and the men of his Indian Defense Association, who opposed the wholesale disposal of Indian lands and the plans to force the tribes up into a false condition of civilization which they could not maintain. Dr. Bland and his followers had been practically outlawed by the orthodox Indian Friends groups and they had little standing in Washington. The officials, under pressure, now made a swift shift in strategy. Only a few weeks back Captain Pratt had flung it in the faces of the chiefs at Lower Brûlé that they would not be permitted to come East until they had mended their ways and accepted the land agreement; now the Sioux agents were ordered to bring delegations to Washington at once. The die was cast. Honest persuasion had failed and, whether they knew it or not, the officials were headed down the road that led to the use of trickery and frontier shakedown methods.

The Sioux had progressed in some ways in twenty years, and the chiefs and headmen who reached Washington in October, 1888, made a better impression than their fathers of the 1865–68 period would have done, if called on to discuss such complicated matters. A few of the chiefs were now able to hold their own with the officials. They spoke sensibly, putting shrewd questions and asking for aid that would assist their people toward self-support. But most of these chiefs were still enough like their fathers to make it impossible for their minds to hold to important issues for any length of time, and their talk flitted erratically from serious to trivial matters in a manner that made the officials, and the Sioux themselves, dizzy. The Indian Office men made an honest effort to keep the talk on the land question, but the Sioux would not have it. To humor them, the harassed officials sat and listened to endless rambling talk, liberally peppered with complaints and demands.

Thus the Pine Ridge delegation brought up the old and very sore matter of payment for the ponies taken from Red Cloud's and Red Leaf's bands by Sheridan's troops in 1876. These claims had rested for years in Indian Office pigeonholes, and it would have been the honest and straight method to inform the Sioux that such matters had nothing to do with a land agreement; but to please the Indians the officials promised to bring the pony claims before Congress at the earliest possible moment and to press for a liberal settlement. That ended the talk of a land agreement. The Sioux from the Missouri River agencies were at once reminded of the thousands of ponies General Terry's troops had taken from them in 1876. True, they had been paid years ago for those ponies, receiving cows and bulls to start them at stock raising; but the officials they were now talking with did not seem to know of that, so the chiefs demanded payment for the ponies and the officials made another pledge, to keep the Sioux in a good humor. By this time the pony claims had mounted up to $200,000, and the herds of ponies had broken the gates, opening the way for a flood of demands based on ancient promises that had not been kept and new promises that the Sioux wished the officials to make, although these officials had no authority to embroider a land agreement with such matters as a nice shiny red farm wagon for every Sioux family, plenty of rice and dried apples in the rations, and money—not money to be put away and to be expended only on official authority for uninteresting things, such as schools and farm tools, but real money that the Sioux would hold in their hands and take to the traders' stores, to buy what they really wanted.

The worried officials, dodging and twisting under this rain of demands, making promises when cornered, getting out of making them when they could, kept bringing the chiefs back to the matter of the land agreement, and a little at a time they collected what they imagined were pledges of support from this chief and that headman. At last the moment came, they thought, for pinning these Sioux down to a definite promise to go home and urge their people to accept the land agreement; but the moment the matter was put into words, the chiefs became indignant. What was this talk of their pledging the tribe to sell any land? This talk here was only about payment for ponies and the red wagons, rice, dried apples, and other nice things the Sioux were to get. An official count exhibited that a small minority of the so-called progressive chiefs stood ready to support the land agreement if certain additional concessions were granted, but the majority of the chiefs refused to commit themselves. They simply broke out into an orgy of fresh demands. Some of them held out hopes to the officials that they might be drawn into some kind of pledge about the land matter, but first they wanted the earth with a three-wire fence around it.[5]

The conferences were broken off and the Sioux were sent home on the first possible trains. What kind of a business transaction was this? The Cleveland administration had been pushed and hustled by the Dakota leaders, the Indian welfare leaders, and now by the Sioux leaders until they hardly knew where they were. They had been led first to deny on the reservation that the chiefs had any authority to speak for the tribe; they had then been urged to eat their own words, bring these chiefs East and accept them as the representatives of the tribe, only to find that it was impossible to deal with them. The avowed purpose of the whole scheme, from the point of view of the Indian welfare groups, was to take the land and pinch the Sioux so severely that they would have to work and support themselves; but if half the demands these Indians were making should be met, the Sioux would not have to do any work for at least two generations. They would all be rich. This dilemma of the Democratic officials did not endure for long. The Sioux chiefs had hardly reached home when the nation voted

[5] The report of these Washington councils with the Sioux is in *Report of the Commissioner of Indian Affairs* (1888), lxxiv. McLaughlin gives some details in his book, and the newspapers of the day contain considerable material on the councils.

(November, 1888), and Benjamin Harrison and his Republicans were swept into power.

In 1889 a benign Providence seemed to be at work arranging everything exactly to the taste of the humanitarians who had set themselves the task of forcing the Sioux to give up their lands and submit to being hustled along a dream road named "progress." The new President, Benjamin Harrison, was a man of their own party with whom they had high hopes of being able to work; but it turned out to be even better than that. Harrison knew nothing about Indians and, in a nice way, cared nothing. The niceness came out when he appointed J. T. Morgan as head of the Indian Office and instructed that gentleman to go to the leaders of the Indian welfare societies and to accept their views as the official Indian policy of the new administration. Never in the history of their long crusade had the humanitarians had such luck. President Harrison had handed them a blank check.

When Morgan came out with the statement of policy, which all new commissioners of Indian affairs considered it their first duty to issue, he stated that it was his "long established conviction" that the Indian reservation system was an anomaly that could not be permitted to persist; the Indians must be made over quickly into imitation whites; the tribes must be destroyed; the Indian youth must be trained for the new life, must be taught English. As soon as possible all the Indians must be absorbed into the body of the white population. The political patronage system must go, and the Indian service must be filled with men and women of a high-minded type and fine integrity. None of this was new. Morgan was repeating the language of the more fervid and radical of the Indian welfare leaders. Before making this statement of policy, he had gone to the Lake Mohonk meeting and told the leaders of the self-styled Indian Friends that he was placing himself entirely in their hands.

Like Cleveland before him, Harrison had made campaign promises that if elected he would not remove men from federal positions for political reasons. He had hardly entered the White House when news seeped out of wholesale dismissals of Democrats from Indian Service positions to make room for Republican appointees. The Indian welfare leaders were all fervent Civil Service reformers, and they had complained loudly when Cleveland had removed Republicans from Indian service positions and replaced them with loyal Democrats. Now they had little to say. They had been given control of the administration's Indian policy and they were too busy making new plans to bother about violations of the Civil Service rules.

They were going straight ahead with their plan for taking nine million acres from the Sioux. The starting point of their plan seems to have been the outrageous assumption that the little group of Sioux chiefs brought to Washington in October, 1888, had made some kind of promise to speak for the land agreement and that it might therefore be represented that the Sioux nation had expressed itself favorably on this matter. The news from the reservation refuted such a view. The Sioux seemed just as strongly against any sale of land as ever, and many of the chiefs who had hesitantly expressed themselves in Washington as favorable to the plan had recanted the moment they got home and had to face the people. The Washington officials knew this; the leading Indian welfare men knew it; and after examining the course of the land commission of 1889, one cannot doubt that some group of men in Washington in the winter of 1888–89 made a careful plan for breaking down Sioux opposition to the agreement and compelling them to approve it. The campaign was planned in minute detail, like a military operation. The officials chose as the chairman of the new land commission the one man in the land whose presence would give the Sioux the strong impression that the army stood back of the land commission. This man was Major General George Crook.

The Sioux chiefs were at work preparing a united opposition to this effort to take their lands when word came that Three Stars —General Crook—was coming. The Sioux grew very thoughtful, and some of the so-called progressives—men who usually followed the advice of the whites—began to draw away from the rest of the tribe. Their sensitive noses had caught the scent of trouble in the air and they were going to be very careful. Three Stars was no small man like that Captain Pratt who had come out with the land agreement in the previous summer; he was a big chief, and when they thought of him, they saw regiment on regiment of cavalry standing in their ranks, ready if he lifted his arm. Rightly or wrongly, the Sioux saw in the coming of Crook a threat. The pretense in Washington that the General was being sent because he was an old friend and a good friend of these Indians hardly deserves comment. Crook had not been as hard on the Sioux in 1876–77 as some other generals, but he had been hard. The Sioux respected him; perhaps they still feared him, few of them called him a friend.

The other two members of this Sioux commission were also big men. They were Charles Foster of Fostoria, Ohio, former governor of that state, and General William Warner, head of the veteran association, the Grand Army of the Republic. This commission of three was given a fund of $25,000, a sum so much larger than the amount needed for ordinary expenses that it could be regarded only as a war chest, and this view has been confirmed in recent years by the publication of General Crook's autobiography. The need for such a war chest had become apparent to the planners in Washington and to the leaders of the Indian welfare groups who were heartily supporting the campaign against the Sioux. The Indians were standing united back of their chiefs and tribal councils in solid opposition to any land sale. The purpose was to smoke them out, to employ a small army of white squawmen and mixed bloods to talk the Sioux dizzy, to induce them to desert their chiefs and sign the land agreement. One thing worried the planners. Dr. Bland and his Indian Defense Association also had a war chest and were already spending money to build up opposition to the land agreement both on the reservation and among the citizens of the Eastern states. Dr. Bland had circulars printed in the Sioux language, urging the Indians to stand fast in their opposition to any land sale, and agents of Bland were handing out these printed leaflets to every Sioux family on the reservation. Bland added to the concern of the Washington officials and the leaders of the orthodox Indian welfare groups by writing stirring personal letters to Red Cloud and other Sioux chiefs, urging them to refuse to even talk to the land commission and promising strong support from the Indian Defense group.

Thus in the early summer of 1889 the battle lines were formed and the Sioux stood in their ranks facing the united forces of the Washington officials and the groups of crusaders who called themselves Indian Friends. The Sioux had no intention of giving in this time, and any true friends of Indians might well have hesitated or drawn back from a struggle that even if it succeeded in beating down Sioux opposition was certain to leave the Indians with a deep conviction that all white people were their enemies. But the leaders of the crusading groups, strong in the belief of the righteousness of their purposes and doubly strong since President Harrison had made their Indian policy his own, had no intention of drawing back. This was a struggle of enlightenment and progress against the forces of darkness and barbarism assembled on the great Sioux reservation, and the crusaders would never call retreat. They would march straight on to victory.

8

Shakedown

THE SIOUX commission went first to Rosebud, in May, 1889. At this agency the Indian Office was striving at great expense to build up the farming operations of the Indians and the commission's arrival coincided with the very important work of seed planting. But the sacred farm program was now looked upon as of less importance than the work of the land commission, and all the Brûlés were ordered to abandon their little farms and come and camp at the agency in tipis in order to attend the land councils. Many of the Indians had to come from points forty miles away, and Swift Bear's Corn Band came from their settlement on the lower Niobrara, about one hundred miles away.

The moment the Sioux arrived and set up their white canvas tipis the land commission reached into its money box and began to purchase beef and other food to provide feasts for the Indians, to put them in a good humor.[1] The Brûlés—eating one big meal after another in rapid sequence—grew happy. They wanted to dance; but the Indian Office now had strict rules against dancing. The chiefs went to see Three Stars (General Crook), who said a word to the agent, and the rule against dancing was temporarily suspended. Dancing having started, other heathen practices, now forbidden, were revived and the Indians threw off all restraint. General Crook and his two fellow commissioners sat in chairs, contentedly watching. The Reverend W. J. Cleveland, Protestant

[1] One of these feasts featured the consumption of fifteen head of beef cattle, paid for by the land commission, to put the Indians "into a receptive frame of mind." *General George Crook: His Autobiography*, ed. Martin F. Schmitt (Norman, 1946), 284.

Episcopal missionary at Rosebud, went about with a look of pain on his honest face, noting this recrudescence of long-forbidden heathen practices. He commented on the fact that even the Lord's Day was being given over to these outrageous dances, and General Crook looking smilingly on.

Crook was being wonderfully nice to the Brûlés. These Indians had counciled among themselves long before the commission came, had put their affairs into the hands of the chiefs and tribal council, and had adopted Dr. Bland's urgent advice not to talk. They ate the lovely feasts provided by the commission's money, danced, and enjoyed themselves, all the while keeping an eye on the land commission to see what it was going to do; but it did nothing. The three commissioners were just good friends who had come to see the Brûlés and to provide feasts and other entertainment. They were most cordial to any chief who cared to visit them; they answered questions when asked. The paid agents of the commission circulated among the mass of Sioux, talking with this influential man and that. Surely, there was no harm in the commissioners. They were good men.

Old Swift Bear was the first victim of the General Crook strategy of simple friendliness. This Corn Band chief had never been able to resist the blandishments of white men. Long ago, before the reservation was thought of, when his band was leading the life of free hunters south of the Platte, he had developed a notable propensity for taking advice from white men. Now in 1889 he was pleased to be seen hobnobbing with General Crook and the other two commissioners; presently he forgot that his people were opposed to the land sale, even against any talk of it. He began to ask questions and received the frankest answers; then he advanced another pace, asking that a council be assembled. Crook agreed.

Some of the talks were in the council hall at the agency; but the Sioux had formed a circle of wagons enclosing a space about one hundred yards across on a grassy level near the agency, and here the big council assembled, wagon seats under a tent-flap shade being placed in the center of the circle for the commissioners, while the Sioux sat on the grass inside the circle of wagons. When all had assembled, General Crook invited Swift Bear to speak, but the old man was in no condition to comply. A victim of the commission's liberality in the way of feasts, he sat with both hands over his extensive mid-section looking very sad. He asked the commission through his interpreter to put off the council. This was done.

On June 4, Swift Bear asked for another meeting, in the face of the tribal opposition, and when the Indians met, he stood up and began to speak boldly of the land sale. But it was apparent at once that, although they wished to please General Crook, he and his followers were suspicious of the true purpose of the agreement. They were afraid of being tricked, and Swift Bear started by asking many questions, particularly about the farms the Sioux were to have under the agreement. General Warner told him that each head of a family would have 320 acres; young single men would have smaller amounts; but then when the old men died the young ones would come into their land. After that, General Crook spoke, telling the Sioux how pleased he and the other gentlemen were to be among them, advising them to talk the land matter over, make up their minds, and come back in two days for another council. He knew that the Brûlés had counciled before the commission came and had decided to sell no land, but he ignored that fact. Swift Bear promised to get every Indian and mixed blood together and bring them in for a general council. The old man had been weakened by talk. He was seeking a cure through more talk. He and same of the other Indian leaders were already in water too deep for wading, and they could not swim. This land business was a simple problem when viewed from a safe distance. The Sioux did not wish to sell, and that was all. Coming close to the matter these Indians were bewildered by a hundred details, all too complicated for them to understand. It would be interesting to know just how the interpreter, Louis Richards, explained to these Brûlés that under the agreement they were giving up about nine million acres and keeping twelve million. The Sioux knew the size of a four-point blanket; but how big was an acre? At Pine Ridge when this matter came up, the chiefs asked the commission to have one of their men go to the flats near the beef corral and stake out one acre and also a Sioux farm of 320 acres, so they could see with their own eyes what was meant.

On June 7, Swift Bear brought seven hundred Sioux men, including twenty-four chiefs, to council. Before the coming of the commission, these Brûlés had chosen Hollow-Horn-Bear, son of old Chief Iron Shell of the Orphan Band, to speak first in council and to tell the white men that the Sioux had no land to sell, but they made the mistake of putting up Yellow Hair, who was older than Hollow-Horn-Bear and a better talker. Instead of stating the tribal decision that the Sioux would sell no land, Yellow Hair got tangled up in a discussion of school money and the prices to be paid for the land the Sioux were being asked to sell. He was a shrewd and experienced Indian, yet he could not understand why the government proposed to sell the Indian land to white buyers at $1.25 an acre for two years, then for 75 cents an acre for two years, and after that at 50 cents an acre, and the commissioners had to explain that this was because the best lands would be sold in the first two years, and after that the price would have to come down to attract buyers for the poorer land.

Swift Bear now introduced Charles P. Jordan, an Indian trader who had lived with the Sioux so many years that he had come to regard them as his people and had married into the tribe. He was what the Christians in the East termed a squawman; but at Rosebud he was known as an honest, upright, and hard-working man. He was also opinionated, and having decided that the Sioux would do best for themselves by selling part of their lands, he now spoke in council with much severity against the Indians who opposed the land sale. He said that they were men who wanted to sit around forever, eating free rations; that they neglected their families and children; and he lumped all opponents of the land agreement together under this head. The Sioux listened to his speech with growing anger. Surely, he knew that the average Sioux could not think for himself in a big matter like this land agreement, yet here he was denouncing them for putting their affairs under the control of the chiefs and tribal council. He seemed to believe that they should let General Crook and the other commissioners think for them. And who was thinking for General Crook? Crook was a soldier, carrying out an assigned duty by talking nicely about this land agreement. He would have performed his duty in the same unquestioning manner if he had been ordered to take two or three regiments of cavalry and dragoon the Sioux until they consented to give up their land.

The position of the fullbloods and their leaders in the tribal council was growing insecure. Their prearranged plan to stand in their ranks, making no move and refusing to talk, had been disrupted by Swift Bear, who just could not keep away from any important white man who came within reach. Standing in their ranks and refusing to talk was all very well, but with Swift Bear and his followers breaking ranks and all this sniping going on from men like Jordan and some of the mixed bloods, many of the weaker-willed fullbloods were beginning to waver. Some squaw-men and mixed bloods were taking the land commission's pay and were getting hold of many of the Brûlés, over whom they had some influence.

The chiefs knew of a matter that was not written down in the fair copy of the proceedings which the commission's secretary wrote out. Swift Bear and certain other progressives were holding private conferences with Crook and the other two commissioners. These progressives were eager to follow the white men's ways, but they were worried over the loud talk of Captain Pratt, Lyman Abbott, and certain other leaders of the Eastern welfare groups, who were advocating tearing up the Sioux treaties and abolishing rations. The progressives wanted assurances from General Crook personally that if they signed this land agreement, their rations would not be endangered, that the treaties would remain in force. There were certain other promises that they desired Crook to make. They talked it over with the commission in private meetings, and Crook put down in writing a pledge covering all these matters and signed it. From that moment the progressives deserted the tribe and plumped for the land sale. They went out and started working on the other Indians, trying to talk them around.

The tribal leaders now acted, bringing forward Hollow-Horn-

Bear to speak for those opposed to the agreement. He began his speech before the commission by giving his credentials, saying that the tribe had held a council before the commission came, had selected twelve men to represent the people, and those twelve had selected him to speak. The tribe did not wish to deal with the commission. It wished to have a general council of all the Sioux assembled and to deal with the commission in that manner. If four men had to decide a matter, he said, and two were here and two far away at another place, they decided differently and it was a split. Only by bringing the Sioux leaders together could a decision fair to all the Indians be reached.

That was all the Brûlé tribe at Rosebud wished to say to the commission; but Crook, as friendly and helpful as ever, now blandly invited any man who wished to speak to stand up. He was ignoring the tribe and inviting the Indians to deal with the commission as individuals. There was a long silence. Not an Indian moved. At last Louis Bordeaux, a mixed blood, stood up. He was the nephew of Swift Bear, that chief's sister having married James Bordeaux, a trader from St. Louis, in the Fort Laramie country sometime in the early 1840's. Bordeaux now proceeded to tweak the tribal council's nose. Every man, he said, had a right to his own opinion, and no group (the tribal council) had the right to decide for everybody. His side (the land commission) had explained the agreement clearly, and how advantageous it was for the Sioux. The commission was out in the open; it now desired, he said angrily, to have a look at the man who was opposing it. Chief Two Strike stood up at once. He was Spotted Tail's old lieutenant, and ever since the murder of that chief he had led Brûlé Band Number Two, who were openly nonprogressive. From his toes up Two Strike was nonprogressive and he was not ashamed of that fact. He said that the tribal council had chosen Hollow-Horn-Bear to speak for it and for the people, which it had a right by age-old custom to do; that Louis Bordeaux knew it, and need not speak hard words. Good Voice, a Wazhazha worthy, glaring at Bordeaux, said that he was a good man who worked hard; that he was as good a progressive as Louis Bordeaux, and that in speaking for the tribe Hollow-Horn-Bear spoke for him. Hollow-Horn-Bear himself now spoke up, accusing Bordeaux angrily of misrepresenting the tribal council by pretending that it was fighting General Crook, when it was only asking that all the Sioux leaders should be brought together for a council. Crook poured oil on the rising waters. The Great Father, he said, did not wish to call a general council, as that would involve neglect of the Sioux farms during the growing season. That was really funny. The moment the commission came to Rosebud the sacred farming program was forgotten, all the people being summoned to the agency, leaving their farms deserted. They had been kept here for over a week, and when any Sioux had told the agent that he opposed the land agreement, did not wish to sign, and wanted to go home to his farm, the agent forbade it and threatened to punish him and his family if he left the agency. Crook went on to say that his feelings had been hurt by the reports being spread about that he had hired Louis Richards to talk to the leading Indians and persuade them to sign.[2] He was pained to find the Sioux in the condition they were now in. "When I left you before," he said, referring to the withdrawal of his troops from the Sioux agencies in 1878, "I expected much good of you, and here after eleven years I come back and find that you have done very little toward civilization. You have been contented to sit down and eat rations the government gives you, thinking the government is always going to keep you. I find that you have to get passes, every Indian that goes away, just like a child. When I was here before I was proud of you. You were full of manhood, and any decision that was required of you, you could give it right away." Yes, Crook had obtained prompt decisions from the Sioux in 1876–77 by dealing exclusively with chiefs and ignoring common Indians. He was now ignoring the chiefs and fishing for the support of the commoners. His memories of 1876–77 were a little askew. Then he had not talked of progress, but of fighting,

[2] *The Word-Carrier*, the mission paper printed in the Sioux language and given to telling the truth, said in its issue for September, 1889, that the Crook commission had hardly reached Rosebud when reports of attempts to bribe Indians in the land matter began to be circulated. Money was given, and promises were made that Indians would be given certain favors if they supported the land agreement.

and Swift Bear, now his best friend, had been ignored. Crook in those days had sought only for scouts to serve him. Why did he throw it in the faces of the Sioux that they had to get passes like children? Were not his own soldiers and officers in the same position? Was it not the army, in 1876, that invented this pass system and pinned it on the Sioux?

When Crook had done, Standing Bear, the only fullblood who had ever run a store in the Sioux country and a notable progressive, got up to speak for two bands: the Loafers and Brûlé Band Number One. These bands now held the balance between Swift Bear and his ultraprogressives with their mixed blood and squaw-man stiffening, and Brûlé Band Number Two and the Wazhazhas, who did not intend to sign under any circumstances. Standing Bear said that his group wished to sign, but that they had dreadful doubts and were hesitating and holding back. Crook forgot for a moment his pose of friendly visitor and spoke with his 1876 voice, sharply, almost threateningly. Then all at once a compact little group of mixed bloods, some of whom were known to be in the commission's pay, started to push their way through the throng of Indians to reach a table on which lay a copy of the agreement with ink and pens, ready for the signers. Wild excitement swept the mass of angry Indians. Hollow-Horn-Bear shouted for those who stood with their tribe to leave, and most of the Brûlés rushed off as if the devil pursued them; but old Swift Bear, ready to die for the whites if need be, ran to the table and seized the pen, shouting to have his name written down. Close behind him came another Brûlé worthy: Crow Dog, the murderer of Spotted Tail. The Sioux who watched began to yell fiercely. The tribal leaders were trying desperately to shepherd the waverers out of reach, before they could be coaxed or bullied into signing, and Agent Spencer had sent his armed police into the mass of Indians to stop this attempt to coerce the weaklings, whose names were needed by the commission. The tribal phalanx was being riddled by desertions. Hollow-Horn-Bear rushed back and begged General Crook to permit the Indians to go home to their farms. Crook soothed him. Why not stay until tomorrow and have another council? Why, indeed? The more talk, the more desertions. Crook told the excited young chief that it was always bad to quit while mad, and that he could see by Hollow-Horn-Bear's eyes, that he was hopping mad. Hollow-Horn-Bear denied it heatedly. If his eyes were red, it was because he had been up all last night studying this land problem. From a man who could not read that seemed a quaint statement; but it was true that the tribal leaders had been up all night, arguing and quarreling. Agent Spencer now came and informed Crook that his police had stopped the trouble outside. Hollow-Horn-Bear, Yellow Hair, and White Horse had tried to chase the Indians away like a flock of birds. He had stopped that. The hot spell was over, and any Indian who wished to sign the paper was now free to do so. They were free men. None of them would be permitted to go home until this business was settled. In effect, all the "no" men were to be kept right there until a sufficient number of them became "yes" men and the land agreement was declared approved.

By this time some three hundred men, led by the mixed bloods and squawmen and nearly all of them belonging to the progressive Loafer and Corn bands, had signed the agreement, and someone was spreading it about that anyone who failed to sign was disloyal. That must have come from Crook, who in all his dealings with the Sioux harped on the word *loyal*. There was now a hint that the disloyal would be punished later on, and the leaders of the opposition began to worry about their future. Hollow-Horn-Bear talked earnestly to Crook concerning his father, Iron Shell, loyal chief of the Orphan Band. Was he not the very first Brûlé chief to sign the treaty of 1868? Had he not led in aiding Three Stars during the Crazy Horse troubles of 1877? One by one the Brûlé chiefs came forward, to protest their loyalty. Crook and his companions sat through endless councils, day after day, listening with amazing patience to all this talk and reaping their reward in a continuous trickle of Sioux coming in to sign the agreement. On June 12 they announced that three-fourths of the adult males had signed and that, as far as Rosebud was concerned, the land agreement had been approved.

This performance at Rosebud closely resembled that put on by the Newton Edmunds commission at Standing Rock in 1882.

Crook's commission, it is true, had not uttered the ugly threats that the Edmunds commission had indulged in; but why need they? In the view of the Sioux, Three Stars was the army. He was right there in front of them, and when he talked of loyalty, they began to ponder how long it would be before he lifted his arm, to signal the troops to march in. Crook was very friendly, with smiles for everyone; but the paid agents of the commission were circulating reports that dire consequences would ensue if the Indians did not sign the paper. Bishop W. H. Hare of the Protestant Episcopal church was at Rosebud during the councils, and he stated in a pamphlet printed by the Indian Rights Association in 1890 that the commission "carried persuasion to the verge of intimidation." That was written by a man who approved of this land agreement, but he did not approve of bullying. The whole program at Rosebud had been to corral the Sioux at the agency and keep them there under pressure until they were worn out and frightened into altering their almost unanimous opposition to the land sale into a three-fourths vote of approval.[3]

Crook had made the fullest use of the squawmen and mixed-blood element, who nearly all favored the agreement, in that they expected to benefit personally under the new conditions it would create. These men could hold their own, living like whites and competing with whites on even terms, and they had none of the fears the fullbloods entertained of disaster that lurked behind the bright curtain which depicted the new day dawning. Crook's commission put forward these squawmen and mixed bloods as the progressive, forward-looking element, the hope of the Sioux; but all the while the Sioux regarded these men as little less than black traitors who were betraying their people. It must be remembered that as late as 1882 no fullblood Sioux would accept a squawman or a mixed blood as his social equal. Only men of pure blood were real Sioux. It shocked the Sioux to see these men speaking in council with an air of being superior to any fullblood, simply because they knew more about the whites and their ways. The Sioux would have had it out with this class there and then, but they were

[3] The journal of the Crook councils at Rosebud forms part of the *Crook Commission Report*, in 51 Cong., 1 sess., *Sen. Exec. Doc. No. 51.*

overawed by the presence of General Crook; and, frightened by the talk of the consequences to ensue to any man who did not sign this paper, many submitted and had their names put down. There had been unity at Rosebud before the Crook commission came. When it left, the Brûlé tribe was split into quarreling factions.

In these land councils Spotted Tail's sons took no part. The family of the great chief had sunk into obscurity only eight years after his death. This was the last "treaty" signed by the old Brûlé leaders. They brought Grand Partisan of the Corn Band from his log cabin on Little Oak Creek east of the agency to let him hold the pen while his name was written down. He said that he was eighty, born in 1809. Swift Bear was now very old and, according to Hollow-Horn-Bear, was growing a bit childish. As we have seen, Two Strike and Red Leaf opposed the agreement. Ring Thunder took no part in the speech making, but let his name be written down. These men were all that were left of the fine chiefs who had led the tribe in the old free days, down along the Platte.

Over to the west of Rosebud, about one hundred miles away at Pine Ridge, another chief was waiting to try his strength against his former opponent, Three Stars. Red Cloud was growing old and his sight was failing; but the excitement over this land agreement had brought to him a St. Martin's summer of contentment; his people had turned to him for leadership, and as head of the tribal council he was in charge of the strategy which he hoped would defeat the land commission. Thus when the commission reached Pine Ridge from Rosebud it was greeted by an imposing array of Oglala tribal soldiers and warriors, armed and mounted and decked in all their war finery. General Crook cocked a disapproving eye at this assembly of fighting men and told the chiefs curtly that he did not like it. He said that the commission had come to deal with the Sioux as individuals, not as members of a tribe controlled by armed Indian soldiers. The chiefs gave a signal and the mounted warriors rode from the field, but Red Cloud and his lieutenants were not as much taken aback by Three Stars' displeasure as the Sioux at Rosebud would have been. The Brûlés of Rosebud had no big chief to lead them; the Oglalas had Red Cloud, and even if he was growing old, his name alone held magic. Moreover, Dr. Bland's Indian Defense Association had concentrated its effort to defeat the land agreement at Pine Ridge. It had filled that reservation with printed leaflets, urging the Sioux to stand united against any sale of land, and it was spending considerable sums in hiring influential mixed bloods and squawmen to work against the agreement. All this gave the Pine Ridge chiefs a feeling that they had strong support among the whites, and when General Crook spoke sharply about the armed warriors Red Cloud and the other chiefs did not flinch but redoubled their efforts to keep the Indian ranks unbroken. They were not going to stand aside and see the lands taken, just because General Crook looked displeased.

The first council with the Pine Ridge Indians was held in the open field near the agency. Several thousand Indians sat in a vast circle, in the center of which the commission, the interpreters, and some of the chiefs sat under a tent-flap shade. The land agreement was read, paragraph by paragraph, the interpreters turning the English text into Sioux. The moment the reading was concluded the chiefs gave a signal and a large body of mounted tribal soldiers rode into sight and began dispersing the assembled Sioux. At this General Crook lost his temper and spoke bluntly to the chiefs; but Red Cloud and the others stood up to him, stating that the tribal council was in control and its orders were that no one was to be permitted to discuss the land agreement, which the tribe had already rejected. The chiefs asked the commission to accept this rejection and to leave the agency at once. Crook ignored the chiefs and went to work on individuals. "Lovely day," he wrote in his diary on June 9. "Tuned different Indians up. Got a good many signatures by different younger Indians who were made to see that they must think for themselves, and in this way it is breaking down the opposition of the old, unreconstructed chiefs."[4]

The General was employing exactly the same method as at Rosebud; sitting with his two fellow commissioners at the agency, smiling, friendly, and endlessly patient, while his squawman and mixed-blood agents brought in the Sioux singly or in groups, to be talked to and persuaded. On June 17 the commission decided

[4] General Crook's *Autobiography*, 286.

that it had sufficient pledges to warrant coming out into the open, and on that day it brought the land-agreement rolls out and invited the Indians to sign. The squawmen and mixed bloods promply started the parade, and with them went a considerable number of fullblood Cheyennes, those Cheyennes whom the Pine Ridge Sioux had befriended during their time of great trouble in 1878–80 and who were now paying off that debt of gratitude by signing a paper that would take half the Sioux lands. The Oglalas were very angry; but they felt helpless, caught in a white man's trap. If they stood up boldly for their rights, Three Stars would call them hostiles, as he was already calling their tribal council an outlaw group. The chiefs, believing that no Sioux could hold his own in a discussion with the commission, had ordered everyone to keep still; but this was working to the advantage of the commission, whose members and paid agents talked steadily, causing many of the Indians to waver in their opposition to signing the paper. This could not be permitted to go on, and the chiefs now put up American Horse to talk for the tribe.

It proved to be a better move than they realized it could be; for American Horse loved to talk, he talked very well, and he talked endlessly, soon proving himself to be the sorest sort of annoyance to Crook and the other commissioners. They were posing as gentlemen of leisure who had come up to visit their Oglala friends at Pine Ridge, with endless time at their disposal for just sitting about and listening to Oglala talk; but this was not at all the true situation, as they must soon go on to the next agency, and while they were at Pine Ridge every moment was supposed to be devoted to business. Now American Horse was taking up their time, and their pretense of abundant leisure made it inexpedient for them to protest. The worst of it was that this chief, the son of old Sitting Bear of the True Oglala Band, had been singled out by General Crook in the old Red Cloud Agency days of 1876–77 as a particular friend, and the General now was in no position to treat him in any but a most friendly way. Besides, American Horse was a notable progressive, and the fact that he was gladly doing all the talking against the agreement was a flat denial of the commission's assertion that the only unreconstructed nonprogressives opposed the land sale.

Day after day, American Horse came to the commission with a big following of admiring Oglalas and requested permission to speak on the subject of the land agreement. Crook and the others, who had a poor opinion of this Oglala's comprehension of the agreement's meaning, had to give a smiling assent; then they sat and listened hour after hour. The next day, just when they were settling down to some real work, this chief would come back, fresh as a daisy and with the usual throng of admiring Oglalas, and put the wilted commission through another half day of oratory.

American Horse had been to Washington on several occasions with Red Cloud and the other chiefs, and he had also toured the entire country with a Wild West show. To the other Oglalas he was a much-traveled man who knew the whites and their ways better than any other chief. And he was an orator. Crook was a poor hand at speech making, while the other commissioners were little better, and the General noted down ruefully in his diary that American Horse "is a better speaker than any of us." He not only spoke effectively, he had a seemingly endless list of subjects to speak on, and he kept wandering off the matter at issue to hold forth on the ration situation, the boundaries of the Oglala lands, the manners and customs of white people, mostly very queer (here the Oglala audience roared with laughter), the much too low pay of Indian policemen, the despicable trickery of Indian traders, and the deplorable conduct of land commissions (not the honorable commission now present, but those that had come among the Oglalas in past years, and had lied, and lied, and lied). Going straight on, this most talkative Oglala dealt with general Indian policy, education, and the scarcity of hard money on the reservation. The commission, which had talked the Brûlés of Rosebud into a dazed condition, now were approaching that state themselves. If this creature would only stick to the land question and get down to the meat; but he would not! From land he hopped to reservation boundaries and took up an hour telling the commission what it already knew better than he did concerning the names of streams and landmarks and the distances all along the border; and as he went tirelessly on boring the commission, Crook and the others noted with wry faces that the Oglala audience was actually admiring this windbag. At last came a day when it was no longer bearable, and forgetting the rule of the commission to never cease smiling and never say a cross word to the Sioux, General Warner lost his temper and broke in upon one of American Horse's endless harangues with the sour remark that he had heard of men being talked to death but had never put much faith in the thing. Now he knew that it was so; but he hoped that their friend American Horse was not going to persist in this attempt to kill them with words. When the interpreter turned this into Sioux, the Oglalas shouted with laughter. Their orator stood in a dignified pose while they laughed; and as soon as he could make himself heard, he went straight on with what he had been saying.

This chief's oratory was the worst hindrance the commission had struck since entering the Sioux reservation; but General Crook was now at work behind the scenes, taking counter measures, and presently American Horse grew hesitant and faltered in his opposition to the agreement. The chiefs went flying to Captain Pollock, the acting agent, and asked for permission for the people to go home. Their cattle and crops needed attention. When the Sioux, who had no love for work, begged for permission to leave the agency, where feasting, dancing and daily oratory were being lavishly provided, because they wanted to go home and work in their fields, something was seriously wrong. In the Crook autobiography we have a hint of what it was. American Horse and some others had been tolled in for private talks with Crook and he had shaken them up by stating that the Sioux could sign this paper and keep part of their lands or refuse to sign and lose all. This was hardly the exact truth, and it sounded like a threat; but it was effective. On June 21, Crook wrote in his diary: "Had a big council this afternoon in which American Horse, Bear Nose and a couple of others made speeches in favor of the bill for the first time since we first met the Indians here. American Horse's band commenced signing. I had coached Bear Nose."[5]

[5] *Ibid.*, 286. And why was Bear Nose so very important to this land commis-

This mighty tussle had lasted for two endless weeks when the break came. If the tribal council could only have locked American Horse up each day after his splendid display of oratory, there might have been no break. Red Cloud and the older chiefs had been right: white men's talk was poison to the Sioux. But Crook and his companions had not gotten at many of the Indians, and, after all, American Horse's little True Oglala Band was not much of a trophy for Three Stars to display. Only one of the older chiefs had been led into putting his name to the paper. No Flesh, old and sick and imagining that he was going to die, wished to make one last friendly gesture toward the whites. He came in and let them put his name down. The other names on the list were almost entirely those of squawmen, mixed bloods, and Northern Cheyennes. Do what he might, Crook could not further break the ranks of the Oglalas after he had managed to cut out the little American Horse group. In his bitterness he claimed that all the opposition at Pine Ridge came from Red Cloud and his old non-progressive following—the people whom Agent McGillycuddy had termed the blatherskites. But this was not true. Man-Afraid-of-His-Horse was always rated as a progressive; but nothing would induce him to have his name set down. Little Wound, Red Cloud's principal rival and a progressive, would not sign. American Horse knew almost at once that Three Stars had landed him on the wrong side of the fence. He claimed that he had been made dizzy by talk that he did not know what he was doing. He said loudly that no one knew better than he did what this agreement meant. It took half of the Sioux land today, and tomorrow the tax man would come and tie strings to every bit of land the Sioux had left. Then the Sioux would have no money to pay the tax man, and he would pull on the strings and drag all the land right out from under the Sioux.

White Cow Man, a Wazhazha, was also wise in his generation. He was in no manner impressed by the promises of the Crook commission of money in large amounts that was to come to the Sioux from the sale of their lands. "I think," he said to Three Stars,

sion? This Indian's name is not mentioned in any other records. He was certainly not an important chief.

"that if I would spread my blanket down here and pile the money that high [indicating four feet], I don't think I could keep it two days. Whenever I get ten dollars, I put it in my blanket and go to any of these traders' stores, and before the day is out I spend it all. I am an Indian and I do not know how to take care of money. Over here at the boarding school [pointing] I have a child that has been there four years. This young one that is at the boarding school, I think he is the one to take land allotments when the time comes." Crook believed that these stubborn Oglalas had a secret reason for opposing the land agreement. *This* was their secret reason. They had not hatched any criminal plot, but were simply frightened at the attempt, for which this land agreement was the starting gun, to rush them suddenly into the ownership of land as citizens and taxpayers. Like this Wazhazha man, the most intelligent Sioux knew that they were unprepared to face such new conditions with any hope of success, and they wanted to wait until the "young ones" were grown men, by which time the Sioux, some of them educated, might be better equipped to hold their own among the whites.

The land commission had been forced to put in a shocking amount of time at Pine Ridge. Its patience was finally exhausted; a last council was called with all the Indians, and General Warner told them curtly that half the adult males had signed and those who had not done so had definitely turned their backs on civilization—which was a quaint way of putting it. Crook then spoke to the Sioux as he had in 1876 on loyalty, loyalty to a government to which they owed none. Many chiefs, he said, had promised not to speak against the agreement, but that was not enough! They should prove their loyalty by signing the agreement and setting their men an example. Evidently this doctrine had no effect; for after Crook left Pine Ridge, he wrote a letter on August 27 disclosing that one of his officers, Captain Pollock, who was still at Pine Ridge, was offering through the Indian agent to supply money to Red Cloud, Young-Man-Afraid-of-His-Horse, and Little Wound, the three principal chiefs. Each chief was to get two hundred dollars to feast his band, if he signed the land agreement. But the chiefs would not sign.[6]

[6] *Ibid.*, 288.

There were said to be 1,366 adult males at Pine Ridge. Of this number the commission while at the agency obtained the votes of 516. Of these, 147 were white squawmen and mixed bloods, 86 were Northern Cheyennes, and only 273 were Sioux fullbloods. After the commission left, Captain Pollock and the agent labored for weeks to obtain additional votes but, even by offering money for feasts, they secured only 158 new names. A three-fourths majority was necessary for approval. Therefore the Oglalas of Pine Ridge had rejected the land agreement.

From Pine Ridge the commission went to the two small agencies on the Missouri: Lower Brûlé and Crow Creek. Under terrific pressure from the land-hungry Dakota whites, the Sioux at these little agencies were demoralized. They feared that all their land would be taken, and they were ready to listen to General Crook, whom they knew by reputation and were inclined to regard as a friend. At Lower Brûlé the big chief was Iron Nation. He informed General Crook that his people feared that under this land agreement the whites would take all their land. The General tried to persuade him that this was not correct, and when persuasion failed, Crook made a promise, evidently in writing, that if the lands of the reservation were taken by the whites, the Lower Brûlé Indians would be resettled on good lands on the Rosebud reservation. Iron Nation then permitted his name to be signed to the agreement, and the other Indians eagerly pressed forward to have their names set down. It was an easy victory, and the commission, being pressed for time, left for Crow Creek, while the Lower Brûlé agent continued to obtain additional signers.

The number of promises and pledges the Crook commission made during these land councils was extraordinary. In a way it amounted to bribery. The Sioux were particularly eager to have General Crook make personal pledges, for many of the chiefs knew him and had strong faith in his honesty. Crook was an upright and conscientious man, very sensitive in matters of honor; yet he made one pledge after another to the Sioux, and he should have known that the officials in Washington would not carry out many of the pledges he made, and that Congress would refuse to vote the funds necessary to execute some of his promises to the

Indians. His pledge to Iron Nation turned out to be an empty one, for all the Lower Brûlé lands were taken by the whites, and when the effort was made to settle the Lower Brûlés at Rosebud, the Brûlés of that reservation refused to give any lands to their Lower Brûlé cousins. The Lower Brûlés then accused General Crook of trickery. In that they were unfair; but they had lost their lands by following the General's advice and they were angry.

Across the Missouri from Lower Brûlé the Crow Creek Sioux were divided, half for and half against the land agreement. This was the smallest of the Sioux reservations, and the opposition to the agreement lay mainly in the fact that so much land would be taken that there would not be enough left for a farm for each Indian family. The commission arrived on July 5 and spent a restful week with few councils. That half of the Sioux who wished to sign did so at once; the other half, led by the two big chiefs, White Ghost and Drifting Goose, had determined to have nothing to do with the land commission. But Crook got them to talk by promising that he would call the attention of Congress to the lack of land at Crow Creek. He made other promises; but the faction opposed to any sale of land refused to sign. Pressed for time, the commission left Crow Creek. Crook was so worried, however, over the stubborn refusal of these Sioux to sign that he induced one of his fellow commissioners, General Warner, to remain at Crow Creek and help the agent in his effort to obtain additional signers.

The commission's next big task was at Cheyenne River Agency, up the Missouri from Crow Creek. Here the first council with the Sioux was held on July 13, twelve chiefs and five hundred Indian men being present. It became apparent at once that all the Sioux at this agency were united in opposition to any land sale; but the commission went calmly ahead with its usual procedure, ordering the Indians kept in camp at the agency until all who could be coaxed into doing so had signed. The usual feasts were provided, to put the Sioux into a good humor, and all the rules against dancing and other heathen practices were temporarily suspended. The Sioux ate the land commission's feasts, danced and enjoyed themselves; but not an Indian came forward to sign the agreement.

Courtesy Bureau of American Ethnology

Two Strike in Washington, D.C., in 1872 as a Member of Spotted Tail's Delegation. Photograph by Alexander Gardner.

American Horse (Wasicu Tasunke). Reprinted, by permission, from Hamilton and Hamilton, The Sioux of the Rosebud, *pl. 229.*

Here, as at the other big agencies, the tribal council was in control, but here half the chiefs were former hostiles of the Sitting Bull camp—half-wild, determined men, out-and-out nonprogressives, opposed to farming and to the land sale. Chief Hump of the Miniconjous led these former hostiles. In 1888 he was the only Sioux chief who still dared to defy the agent by maintaining a permanent soldiers' lodge in his camp, to control his people and flout the police. The agent, anxious to gain at least an appearance of control over his Indians, had hit on the expedient of appointing this openly defiant chief as head of the police; so Hump now held the police in one hand and his Indian soldiers in the other. He was the big man at Cheyenne River.

Crook knew before he came to this agency that he would have some very hard nuts to crack, and he made his arrangements. His first move was to order Major G. M. Randall to accompany the commission in full uniform, as a hint to the Sioux that they had better not go too far. The land agreement held out many promises of aid in farming; but at this agency the very name of farming was like a red rag to a bull. The former hostiles did not care to work at all, and the tamer Sioux who had been at the agency for many years had been so slave-driven that they were disillusioned. They had a right to be. Every season the agent and his men drove them to plant fields, and generally in midsummer drought and hordes of grasshoppers destroyed their crops. The next year the hopeful agent would force them to plant more acres than ever before, and their crops would be flooded out. Moreover, the agreement provided that all the Cheyenne River lands south of that stream were to be given to the whites, while these Indians were to be removed to a district farther north where there was little good land, little water, and much land spoiled by gravel. There they would be forced by their agent to attempt to win self-support by farming. The Sioux bands that had been at this agency for years were violently opposed to being removed north of Cheyenne River, and they were determined not to sign the agreement.

This was the only Sioux agency at which the land commission failed to recruit at once on its arrival the support of nearly all the squawmen and mixed bloods. At Cheyenne River the men of these classes were strongly opposed to the land agreement, for the very good reason that they feared it would ruin them if its terms were carried out. They told General Crook that the lands of the new reservation north of Cheyenne River were so bad that a fair crop once in two years would be impossible there. The only lands fit to farm were in stream valleys, and there was not enough such land to provide little farms for half the Cheyenne River people. They wanted Crook to promise that the boundary line set down in the agreement should be shifted farther south, so that the Indians could have the use of the water of Cheyenne River and the good farmlands in its valley, and they thought it would be folly not to change the boundary so as to keep for Indian use the agency buildings, boarding school, and church buildings, which were all south of Cheyenne River. Crook was very sympathetic; but this so-called agreement had been written in Washington to suit exactly the wishes of white men, and his commission was powerless to alter one word in the document. He and the other commissioners, however, made many promises. Still, most of the squawmen and mixed bloods held back. The puzzled commission, sorely needing the services of these men, both in coaxing the fullbloods around and in forming a phalanx that at the right moment would march up to the table and start the signing off with a flourish, now discovered that a sizable group of squawmen and mixed bloods might be won over if certain promises were made, particularly a promise to protect them from Chief Hump and his Indian soldiers.

These were the men of the Bad River colony. Wishing to farm and to raise stock, they had found it advantageous to move to Bad River, in the southern edge of the Cheyenne River reservation, to get away from the troublesome Indians, part of whom regarded their farming as a bad example which should be stopped by threats, while other Indians thought it was nice to go and stay with these men's families and eat them out of house and home. By moving to Bad River the families had escaped these inconveniences, and if the land agreement went through, they expected that many white settlers would move in between them and the Cheyenne River Indians, giving them the isolation they desired.

But they were afraid of Hump. Their families were all on the rolls at the agency, drawing free rations, clothing and other supplies; Hump insisted that they were part of the tribe and must obey his orders.

General Crook now made some promises to these men and won them over. He at once called a general council, and when it met ignored the fact that all of the Sioux were opposed to the agreement. He let the Indians talk, and when the moment seemed right announced blandly that any man who wished to sign the agreement should now come to the table. At once a compact group of squawmen and mixed bloods came forward. The Sioux let out a roar of protest; Chief Hump threw up an arm, and at the signal two naked warriors, painted and with war clubs grasped in their fists, jumped in through the open windows and made for the group of squawmen and mixed bloods who were preparing to sign the agreement. At the same instant more of Hump's tribal soldiers attempted to enter through the door, and even the Indian police, under Hump's orders, joined in the movement. A scene of wild riot ensued, the Indian soldiers and part of the police making a hard push to break up the council and drive everyone away, while General Crook, Major Randall, and Agent McChesney, aided by part of the Indian police and a few Sioux volunteers, strove to restore order and resume the council. A large body of angry Indians, shouting that this agreement was an attempt to steal their lands, tried to fight their way to the table where the hated document was laid out, ready for signing; but Three Stars, grim-faced and threatening, stood in their path, backed by his little group of supporters, and even in the fury they were in, no Sioux dared lay a hand on the General. Crook was no longer posing as the friend of all the Sioux, smiling and endlessly patient. He ordered the signing to start and grimly stood guard while the squawmen and mixed bloods filed up to the table and set down their names. The fullbloods, yelling fiercely, stood looking on. The General had warned them through an interpreter that if they attempted any further violence, he would summon the troops that stood ready at Fort Bennett, within sight of the council room. Crook was thoroughly aroused and in such a temper that he forgot for the moment that he was the dignified chairman of an august commission.

By July 22, after another week of heavy labor, the commission thought that it had collected enough power to smoke the stubborn fullbloods out, and on this day it called another general council; but the Sioux put up Chief White Swan to speak for them, and although this man was utterly friendly and a well-known progressive, he said no word in favor of the land agreement, confining his talk to an attack on the squawmen and mixed bloods for their desertion of the tribe in its time of trouble. When his speech was ended, there was a long and tense silence; then General Crook began to speak heavily on his favorite theme: loyalty. He said that the chiefs were shirking their plain duty to set their men an example by going up to the table and signing. A chief got up to reply to this; the Sioux all began to shout; and then Hump's policemen and Indian soldiers made a sudden push and broke up the council.

The next day Crook called the Indians for a final talk. He was in a very grim humor, and his words were ominous. He stated that the commission was leaving, that he was displeased with yesterday's spectacle of disloyal policemen and the hostiles of 1876 chasing friendly Indians away from a council with arms in their hands. He had learned of threats being made against the lives of the squawmen and mixed bloods who had signed, and he would see to it that these men were protected. The commission was leaving copies of the agreement with Agent McChesney, so that any Indian who wished to sign might do so, and Major Randall of the Twenty-third Infantry (this officer had sat hard on the necks of the Cheyenne River Sioux in the bad days of 1876) was staying at the agency to assist. No Sioux would be permitted to leave, to return to his farm, until the agent was satisfied with the signing and gave his consent.

This was Crook's idea of leaving the Sioux free to vote as they chose. He was simply corraling them at the agency with an army officer whom they feared in an effort to break their opposition to signing. There was also a covert threat that, if needed, troops would be called in. Yet Crook was acting strictly within his instructions to treat the Sioux as individuals. The officials viewed the tribe and the tribal council as outlaw groups which had no right to represent the Indians, and certainly had no authority to use Indian soldiers to enforce their orders. The treaty recognized that tribal authority, but it did not suit the book the officials were now using, and although they had no legal right to do so, they treated the tribal organization as a group arbitrarily attempting to restrict the freedom of individual Indians. These same officials saw nothing wrong in their own order, to coop the Sioux up at the agencies and keep them there until they signed an agreement which they did not wish to touch.

The wisdom of Crook in removing his land commission from the riot at Cheyenne River and leaving Agent McChesney and Major Randall free to deal with the recalcitrant Sioux was soon demonstrated. A kind of blackout curtain was drawn about this agency, and when it was drawn aside two weeks later, the announcement was made that the Sioux here had triumphantly approved the land agreement. There were said to be 750 adult males at Cheyenne River. While at the agency, the commission, although it employed every means to bring the Indians around, had signed only 300, most of whom were white squawmen and mixed bloods. In two weeks, Agent McChesney and Major Randall had run the total of signers up to 620, by methods left unrecorded. It would be interesting to know just what they did to Chief Hump and his embattled tribal soldiers to stop their activities, so that pressure might be brought on the Sioux. The methods employed were very effective. Even Chief Hump signed!

At Standing Rock ever since 1878 the agents had pointed out Chief John Grass as the best progressive they had, the man who led in farming and self-support and was always on the side of the whites. Now, in 1889, Grass stood at the head of a united tribe, determined to sell no land and pledged not even to discuss the subject with the commission. Crook went to work on these Indians on July 26. The Sioux listened to the commission's opening addresses and were polite but very firmly opposed to dealing. The commission hired squawmen and mixed bloods as agents and went after the Sioux, and by the twenty-ninth Chief Grass was enough weakened to make the fatal error of starting a discussion with the commissioners. He said, to begin with, that whatever his decision was, the Sioux of Standing Rock would follow his lead. The tribe, he said, was united in opposition to any land sale. East of the Missouri was plenty of land, open to white settlement. Let them take those lands. This land agreement, he added, had nothing in it that appealed to sensible Sioux, and the commission was not acting fairly. It was not fair to leave a copy of this paper with each agent with orders to get more names signed after the commission had gone away. He had heard that at the other agencies certain white men and mixed bloods were being paid, so much a name, for bringing in signers. If such things were proper, there had been no need for the commission to come among the Sioux. They could have stayed at home and sent out copies to the agents, saying, "Get it signed." At this point the Sioux applauded. Grass went on to state that he knew that the commissioners were honorable men, but that he did not think that they were doing right in leaving copies of this paper with men whom the Sioux did not trust.

This was a good speech; but once Chief Grass had broken silence the Sioux were lost. They attended one council after another, and presently little groups were coming in for private talks with the commission. Agent McLaughlin employed all his influence to bring the leading Indians around, and most of the squawmen and mixed bloods were openly working for the commission. To bring pressure on the fullbloods, the commission now announced that enough Sioux had signed the agreement at the other agencies to insure its approval, whether any Indians at Standing Rock signed or not. This statement, which was not true, worried many of the Indians. They feared that if the agreement was approved without their names being put down they would be discriminated against and would receive none of the benefits accruing to signers of the paper. Chief Grass was frightened and wavering; and here we obtain from Agent McLaughlin's book, written years later, a most interesting item that was left out of the land commission's report. McLaughlin tells us in his book how he slipped over to Chief Grass's house in the dead of night and convinced the chief that if he and his people did not sign, Congress would take all their land and they would get no compensation. It was after this, on August 3, that the commission decided that the time was ripe for a final council and the signing up of the Sioux.

Agent McLaughlin had feared all this time that Sitting Bull and his former hostiles might appear at the agency and attempt to break up the councils. He had his Indian police on guard, and a large force of armed Yanktonais progressives who favored the agreement was being held in readiness; but Sitting Bull's camps were very far from the agency, down on Grand River, and none of the men from those camps came to the councils. On August 3, just as the final council was getting under way, Sitting Bull appeared suddenly with a large number of his warriors. He said angrily that he had not been informed of these councils. Agent McLaughlin denied this. Sitting Bull then demanded the right to address the council; but before he could go any farther, McLaughlin seized Chief Grass by the arm and rushed him to the table, to begin the signing. Other signers crowded up to the table. Sitting Bull's men began to whoop fiercely and jumped on their waiting ponies. McLaughlin snatched up the copy of the land agreement, to thwart any attempt to destroy it, while his police and armed Yanktonais went vigorously to work on Sitting Bull's warriors, pressing them back and out of the way. That done, McLaughlin resumed leading the waverers up to the table and holding them there while their names were written down. Thus, with furious yells rending the air and Winchesters being flourished, the Standing Rock opposition broke down and the Indians were hustled to the table to watch while their names were set down.

The commission now was certain that it had obtained the assent of three-fourths of the adult Sioux males and that its labors were concluded. It left Standing Rock and took the Northwestern train for Chicago. On the train a careful check of the number of Indian names signed to the agreement was made, and it was found that, even with the additional names obtained at Standing Rock, there was not the necessary three-fourths. A message was therefore dispatched from St. Paul to Agent McLaughlin, urging him to obtain a certain number of additional names, and this agent at once set to work on the assignment. With the commission out of the way and the solid opposition of the tribe broken to pieces, he had little difficulty, and soon was able to send the commission more names than they required.

The commission now announced that it had 130 votes more than the required three-fourths majority. Perhaps it had. No one at the time questioned its method of figuring, nor was inquiry made into the tinkering with Sioux population figures which on orders from Washington had been undertaken before the commission entered the reservation, an operation clearly intended to cut down the number of Sioux adult males and render the commission's work of obtaining a three-fourths majority easier. Dr. Bland, Red Cloud's best friend, commented angrily on the methods the commission had employed, which were enough to throw out a dozen elections on the grounds of fraud, misconduct, intimidation, and bribery; but no one in the East seemed to credit such reports. The Sioux reservation when necessary could be turned into a good soundproof room, and if any muffled rumblings of contention had reached the high-minded Washington officials and the equally high-minded leaders of the Indian Friend groups these gentlemen refused to credit their own ears. Their view of it was that the Sioux had listened to the good advice of their old friend General Crook, with good will and friendliness. Thus the way had been opened for the Eastern Indian welfare groups to put into operation their program for giving the Sioux what they were pleased to term a better life. They were highly gratified, and they congratulated the Crook commission on the service it had rendered to the public and to the Sioux nation.

On the great reservation the Sioux, dazed at first from the drubbing the land commission had given them, were beginning to recover from the shock and to look about them, half frightened and half angry, as they tried to puzzle out how their united stand against the land sale had been turned into a three-fourths vote in its favor. Presently some of the chiefs began to cry fraud, and the angry Indians took up the cry; but no one outside the reservation seemed to hear.

9

Messiah

THE LAND commission of 1889 had made so many promises to the Sioux to induce them to vote in favor of the agreement that it was considered most necessary to bring the chiefs to Washington and to get them to approve a digest of promises prepared from the mass of matter the commission had written down in the field. Most of these promises were outside and in addition to the terms of the agreement, and many of them would require special legislation and the voting of money by Congress. General Crook was in no humor for seeing his pledges to the Indians put away in pigeonholes in Washington. He believed that his honor was involved, and when the chiefs were brought East in December, he and the other two members of the commission came to attend personally to the ironing out of differences with the Indians.

The officials were still keeping up the farcical pretense that the Sioux chiefs on the reservation did not represent the tribe but that in Washington they did. All through the long and angry councils during the summer the Crook commission had turned its back on the chiefs, insisting on dealing with the Sioux as individuals; but now in December the commission met the same chiefs in Washington and blandly accepted them as the authorized representatives of the Sioux. During these councils in Washington the Sioux themselves tore apart the official pretense that chiefs were tyrants who did not represent the common Indians. These Sioux said that the chiefs who had first brought the tribe to the reservation were now growing old and some were childish; the Sioux must have chiefs to think and act for them, and they begged the officials to discard the rule that no new chiefs should be recognized and permit the Sioux to select a group of younger and more vigorous leaders to represent them. The startled officials exchanged blank looks. They all knew that President Harrison had taken the policy of the Indian welfare groups as his own, and a leading objective of those groups was to destroy all chiefs and thus to free the Sioux, free them so that they could be controlled absolutely by government-appointed white agents.

General Crook was deeply concerned over the question of beef for the Sioux. At Rosebud the Indians had refused to consider the land agreement until he gave them in writing a pledge that this agreement did not affect in any way their right to treaty to free rations, clothing, and other supplies. He had made similar pledges at the other agencies; yet these Indians had hardly finished signing the agreement when Commissioner Morgan ordered a sudden cut of two million pounds in the beef issue at Rosebud, one million at Pine Ridge, and similar cuts at the other agencies. This was done, as Morgan later stated, because "some hocus pocus" had been uncovered at these agencies, the Sioux having padded their population figures to obtain more rations and other supplies. Thus at Rosebud the Indians were drawing rations on some two thousand dead names, while the Pine Ridge people were doing the same thing on a somewhat less handsome scale. The government, it appeared, was being swindled, and to a man of Morgan's type here was a duty to be performed in the name of righteousness, and he performed that duty at once, ignoring everything except the allegation that the Sioux were cheating the government. Even that supposed fact was only true from a narrow view of the matter. The officials, with no right to do so, had been slashing the Sioux rations until in the early eighties the amount the Indians received was only two-thirds of the full ration due under the treaty. By failing to report deaths and by borrowing children from neighbors to increase the number of persons in a family, the Sioux got around the cuts and managed to obtain just about full treaty rations by padding their population figures. Yes, they cheated; they had to; for if they had simply protested the cuts in their rations, no one would have heard and nothing would have been done to aid them. General Crook's view of the matter was largely personal. He had pledged the Sioux that if they signed the land agreement their rations would not be cut; his honor was concerned, and he protested strongly against Commissioner Morgan's sudden cut in the beef ration. Beef was the mainstay of these meat-eating Indians, and this cut would work great hardship. These Indians had lost most of their crops for two years past, and in 1889 they had been ordered to desert their farms and come to the agencies, to spend weeks at councils with the land commission. That had not improved their chances for a crop, and then drought had come and wrought havoc. The Sioux were already suffering, and even from the point of view of the government's own interest it was vital to keep them contented, now that they had been hustled into surrendering their lands and were faced with a demand that they go to work and win self-support. To starve them at the start was no method for beginning that program with any hopes of success. General Crook's protests were in vain. Commissioner Morgan insisted that these Indians had cheated the government and that only a special act of Congress could restore their beef rations to the former level. There was no other way; yet, when the Sioux crisis broke some months later Morgan found a way to increase the beef ration without authority from Congress—he found the way, when it was too late to help.

General Crook had to be content with this promise that Congress would be asked to increase the Sioux beef ration. He handed this promise to the Sioux chiefs in Washington and the chiefs went home, to feed their hungry people on promises. Gloom settled over the reservation. The Sioux had no faith in Congress, and they were right, for that body of comfortable and well-fed gentlemen dawdled with the question of more food for the Indians, taking the matter up when nothing else required attention, and then laying it aside for future consideration. In the meantime the Sioux were going through a Dakota winter of cold, hunger, and disillusionment.

Congress, while holding up the measure to provide more beef for the Indians, gave prompt approval to the Sioux land agreement, which was to benefit the Dakota whites; and thus after ten years of scheming and striving nearly one-half of the Sioux lands were thrown open to settlement. But there seemed to be a curse on this plan for forcing the Sioux to give up their land; and now that the prize was won it turned to dust like Dead Sea fruit in the hands of the Dakota land boomers. Drought had cured Dakota, at least for the time being, of her chronic land hunger. The year 1887 had been a trying one, with poor crops; then in 1888 the drought increased, and even the good farming districts in eastern Dakota had crop failures. With their incurable hopefulness for better fortune next season, the pioneer farmers of Dakota tightened their belts and hung on, to plant more acres than ever before in the spring of 1889. Crops looked wonderfully good in early summer; then in August the drought struck with savage force. In a land inured to drought, this was the worst on record, and thousands of farmers abandoned their land and left Dakota. The drought not only took the crops, it destroyed much of the pasturage. It was "a thin-grass year" with hay grass hard to obtain, and even the hardy native buffalo grass was burnt a dark reddish brown by midsummer. Land prices fell to record lows, quarter-sections selling for as little as two hundred dollars. The largest farm-mortgage company failed, and the public in the East (which had much of its savings invested in such companies) learned with a shock the appalling extent of the loans that had been made with its money on alleged farms in Dakota and other Western states. There was a class of homesteader who never intended to farm. These men staked out claims, built little shacks, and put up a few rods of fence, to establish their rights as homesteaders. They lived on their land as little as possible, made no improvements, and the moment they obtained a deed from the government they went to the loan company's agent and obtained a loan of $800 to $1,200 by mortgaging their alleged farms. They then vanished, to repeat the performance elsewhere; and when the loan company's agent came around later to inspect the farms he had to foreclose on, he found nothing but drought-stricken prairie, for the neighbors had generally carried off the shacks and token fences, which were the only improvements.[1]

[1] S. K. Humphrey, *Following the Prairie Frontier*, (Minneapolis, 1931), 153.

The backbone of Dakota was its farm families: poor people, some American, but mostly European peasants who had come straight from their homelands to this widely advertised land of plenty, bringing very little with them beyond their strong bodies and stout hearts. They had now lost everything from drought and were worse off than Sioux Indians, who were assured of food, clothing, and other needs through their treaties with the government. Where the Sioux complained of short rations, many of the white settlers were face to face with downright starvation. Back in the 1860's when drought had created a similar crisis in Dakota, the territorial officials had toured the Eastern states soliciting aid for the starving people. It was now proposed that a similar caravan of officials should tour the East; but in 1889 South Dakota was entering the union, and leading citizens denounced this plan for heralding the entry of their state with the spectacle of "bands of beggars" going from city to city asking for help and, in effect, admitting that Dakota was a country in which hard-working farmers could not make a living. The plan was therefore dropped.[2]

And this was why the Dakota whites did not shout with joy when some nine million acres of Sioux land were proffered to them in the autumn of 1889. No one wanted land. Even the most blatant of the land boomers had been temporarily silenced by the disaster the drought had wrought. Far-seeing leaders in the new state were pleased that a way had now been opened across the Sioux reservation for the railroads to advance from the Missouri and connect the Black Hills country with the eastern section of South Dakota; but they knew that the building of the railroads and the settling of the Indian lands would have to wait for better times. The Eastern Indian welfare leaders continued their celebrations over the success of the land commission. They seemed unaware that anything was wrong in Dakota, and they were eagerly waiting for the money to come pouring in from the sale of the Sioux lands to the Western settlers. With that money they were going to boom Sioux farming, build new schools, and hustle the

[2] As the subject of drought is a rather delicate one in Dakota even today, it may be as well to state here that this brief account of the drought conditions in 1887–89 is made up from articles written by Dakota citizens and printed in *South Dakota Historical Collections,* Vol. IX and XIII.

whole Sioux nation onward into civilization. But no money was coming in. A few frontier gamblers bought likely townsites in the Indian lands beyond the Missouri and squatted on them, living on a thin diet of hopes for the day when the railroads would come in and a new land boom would start. That was about all that happened. The vast plains that had been wrested from the unwilling Indians lay empty and silent with most of the sparse vegetation dying from drought. "Brother Dawes Land" the Western newspapers called it with wry humor, honoring Senator Dawes who had fathered the Sioux land bill.

The drought was worse on the Sioux reservation than in the settled districts east of the Missouri; but the Indians were doing little farming and they did not suffer as much from crop losses as the white farmers. It was the shock the land commission's activities had given them and the cut in their beef ration that distressed the Sioux. The Indians did not understand clearly the agreement they had signed; but they had a feeling of intense apprehension that this document put them at the mercy of the whites, who intended to uproot them and cast them into a new and strange manner of living which would be very hard. The Protestant mission society reported the joy of the missionaries over the signing of the agreement and went on to state that this success had thrown the Indians into a dreadful state of mind, "one of uncertainty and almost of consternation, like that of men on a vast ice-floe which is about to break up."[3] Bishop Hare thought that this state of the Indian mind opened up a great opportunity either for good or evil, and it was his prayer that he and his helpers might have the strength to do what they were called on now to do, namely, to lead the Sioux onward into a new life, the very thought of which threw these Indians into the depths of despair. Truly, the Indian welfare groups in their eagerness to thrust the Sioux swiftly into civilization had opened a dolorous road for these Indians to tread.

The dazed mental condition in which the land commission left the Sioux soon began to wear off and the Indians fell to quarreling among themselves, the tribal leaders violently attacking the squawmen, mixed bloods, and Christian fullbloods, whom they accused

[3] *Report of the Board of Indian Commissioners* (1889), 51.

of betraying the people by helping the white men to break down the opposition to any sale of land. At Pine Ridge a feud had broken out between the fullbloods led by the nonprogressive Red Cloud and Big Road, the former hostile, against Young-Man-Afraid-of-His-Horse and old Little Wound (McGillycuddy's two prize progressive chiefs) and a heterogeneous group led by American Horse and made up largely of squawmen and mixed bloods with a goodly stiffening of Northern Cheyennes. The Red Cloud group furiously accused the American Horse minority of betraying the people. They said that the Crook commission had deceived them, and having gotten the lands they wanted, the white men would now forget all the fine promises they had made and would laugh at the Sioux. At this moment the agent received an order from Washington to cut the beef ration, and the angry Sioux gathered in crowds, to jeer the men who had signed the land agreement. Agent Gallagher reported that if this beef cut had come while the land commission was at the agency the commission would not have obtained one vote. Now every man who had signed was being called a fool and a dupe.

The news that a delegation of chiefs was to be taken to Washington, to meet Three Stars and the other land commissioners and arrange for the carrying out of the promises made to the Sioux, strengthened the faltering progressive group at Pine Ridge. Their own man, American Horse, led the delegation; but when the party came home with nothing more than a new set of promises to show for their long journey, the Pine Ridge Indians became very gloomy and "a feeling of indifference about the future" appeared among them. They felt trapped. Winter was at hand and the beef cut, taken in connection with the loss of all crops, had produced a state of semistarvation. This was nothing new. Crop failures and the cutting down in the appropriations for rations by Congress had caused suffering at Pine Ridge every winter from 1886 on; but now, in December, 1889, there was less food than ever and the mental depression widely spread among the people caused many of the Indians to lose heart completely. They said that they did not care what happened to them now. Their children were dying and they might as well die, too. Bishop Hare stated that an epidemic of measles swept over the reservation in 1888–89, followed by a terrible attack of grippe in 1889–90, and the children died by hundreds from whooping cough. To white people measles seemed a small matter and whooping cough was nothing specially bad, but among the Indians these childish complaints assumed a deadly form, and they often attacked adults and killed them.

The Sioux at Rosebud seem to have suffered even more in the winter of 1889–90 than their neighbors at Pine Ridge. At Rosebud the beef cut was two million pounds as compared with one million at Pine Ridge, and the Rosebud Indians had also lost all their crops in the drought. They did not have many stock cattle. At Pine Ridge there were about ten thousand head of such cattle, placed in the hands of the Indians to start them in the business of cattle growing. Rules forbade the Sioux killing stock cattle except on a written permit from the agent, and these rules were so sternly enforced that there had been few instances of the illicit slaughtering of stock cattle; but now Agent Gallagher reported that his Indians illegally consumed 700,000 pounds of stock cattle between June 30, 1889, and June 30, 1890. Yet most of his Sioux were still suffering severely for want of food. Perhaps his figures were too high. Again, not all his Indians but only certain bands were indulging in this illicit killing of stock cattle. The agent and his police could not stop them. This breakdown in respect for authority began to be noted early in the winter and was undoubtedly one result of the feeling of despair whose origin lay in the mental shock the Sioux suffered as an aftermath of the land commission's activities.

All through the year 1890, up to the Wounded Knee disaster, that high-minded and stiff-backed educator, Commissioner of Indian Affairs Morgan, stoutly denied that the Sioux had suffered from hunger during the winter of 1889 and the spring and summer of 1890. He proved his case, by quoting figures on beef and other rations purchased for the Sioux. *Figures!* With Indian Office figures you could prove anything. The Sioux agents naturally supported the view set forth by their superior officer, up to the time of crisis in 1890, when some of them were in such danger that they deserted Morgan's cause and admitted frankly that their

Sioux had been very hungry for at least a year. Most of the Sioux leaders complained that their people had suffered greatly from hunger. Such men as American Horse, Young-Man-Afraid-of-His-Horse, Sword (the captain of police at Pine Ridge), and Hollow-Horn-Bear of Rosebud, were not given to lying. They were all rated as progressives who supported the government in ordinary times, and when they said that their people were suffering, it meant just that. Bishop W. H. Hare of the Protestant Episcopal church, who visited the reservation regularly and had known these Indians intimately for years, stated that they had been long in a state of semistarvation and had been cast into deep despair over the frightful death rate among their children.

One missionary and reservation teacher after another told the same story. Commissioner Morgan went on quoting figures to show how much beef had been bought for the Sioux, leaving out such minor facts as that the government contracts called for northern ranch beef while, to save a little money, he had accepted Texas trail cattle. In a drought year with grazing very poor these animals had been driven all the way up from Texas; they were delivered on the Sioux reservation in the fall of 1889 in very poor condition, and they continued to lose weight during the winter and early spring. Captain J. H. Hurst, Twelfth Infantry, an officer stationed at Cheyenne River Agency and who had had long experience among the Sioux, stated that a Texas steer delivered at a weight of 1,200 pounds in November and wintered on the open range would lose about 50 per cent in weight by spring and would then produce about 600 pounds of very poor beef. Former agent McGillycuddy of Pine Ridge said in 1890 that the Sioux agents did not have the steers weighed when they were driven out of the corral on issue days. They usually reckoned one steer for thirty Indians, to last them a week or ten days, and whether the animal was issued in November when in good shape or in March when it had lost most of its flesh it was full ration for thirty Indians. After he became alarmed for himself and deserted Commissioner Morgan, Agent Gallagher of Pine Ridge reported that in the spring of 1890 his monthly beef issue was 205,000 pounds while the full ration legalized by treaty would have been 470,400 pounds. But what did he mean exactly by 205,000 pounds? Was that the weight of the cattle when turned over to him in the preceding November, or had he made fair allowance for the shrinkage in weight? No one knew. And that was why Indian Office figures could be made to prove anything.

Indian reservations were strange places in the 1880's. Partly through a censorship exercised by the agents and partly because the whites beyond the reservation bounds were not at all interested, hardly a whisper of what went on reached the outer world. From August, 1889, to August, 1890, some twenty thousand Sioux, including women and little children, were suffering severely for want of food and medical care, mainly because high-minded Christian men in Washington and the East were too high minded to go behind a set of official figures. The cut in rations certainly was a contributive cause of the outbreak of epidemic diseases. Captain Sword of the Pine Ridge police told the Reverend W. J. Cleveland of Rosebud that in the winter of 1889–90 the Pine Ridge death rate was 25 to 45 a month in a population of some 5,500, most of the victims being small children. Neither Sword nor Cleveland had any method for telling the world of this. As for the agents, they did not consider it good taste to refer to such matters, until forced to do so. And here was one main cause of the despondency that spread among the Sioux during the winter of 1889–90. They were starving and their children were dying, but no one out in the white men's land seemed to know it or to care. Some officials in Washington had done this to them by simply writing an order for a beef cut; and when these Indians thought of the land agreement and of the power it gave to the officials to work their will on them with all the world outside ignorant of what was going on on the reservation, they lost faith in their future. What did it profit them that there were thousands of people in the East who called themselves friends of the Indians? These self-appointed big brothers were happy over General Crook's success in persuading the Sioux to sell their land. They knew nothing of the drought that had taken the Sioux crops, nothing of the beef cut, nothing of the Indian children dying by scores and hundreds. They were happily making plans for a bright future for the Sioux. "The whole Sioux tribe," one of their leaders stated, "must perforce be jostled from the apathy and sluggishness of its old condition and be thrust into one that must, of necessity, compel a struggle in which all will be tested and many saved."[4] Fine talk for philanthropists and idealists to be putting out! It seems unbelievable, but it is perfectly true, that the leaders of these Indian welfare groups and the officials in Washington, after the destruction of the Sioux crops by drought in 1887, 1888, and 1889, seemed unaware of that fact and had the brightest hopes for starting these Indians into self-support from farming in 1890. What particularly pleased them was that the new laws gave them the power to force the Sioux to farm.

Only blind fate would have worked out the grimly humorous scheme of tossing the ghost dance mania among the troubled Sioux at this moment; but fate seemed to be always lurking just around the corner, ready to upset the beautifully prepared plans of the Indian welfare leaders and Washington officials. William Selwyn, a mixed blood who was postmaster at Pine Ridge, reported that the first rumors of an Indian messiah in the far west reached the Sioux in letters from the Shoshoni reservation in Wyoming, late in 1889. Most of the letters were in English, and he read and translated them to the Sioux to whom they were addressed. He was therefore in a unique position to observe the origin of the messiah reports. The Sioux did not know the country west of the Shoshoni reservation in Wyoming, and their best information was to the effect that the messiah dwelt beyond The-People-Who-Wear-Rabbitskin-Blankets who lived west of the Yellow Faces.[5] The messiah had come to save the Indians from the new slavery the whites had put upon them; there was to be a return to the old free life of hunting, and the Indians would be as happy as they had been before the first whites came among them. Some of the Pine Ridge Sioux, who were extremely discouraged and had lost all hope for the future, seized on the rumor of the coming of this strange savior; the talk spread over the reservation, and presently a council was assembled to discuss the news of the messiah and to

[4] *Ibid.*

[5] Unidentified tribes in the desert country of Utah and Nevada.

consider the proper action to be taken. It was later stated by the angry officials that Red Cloud and the reactionaries, who wished to block all efforts toward progress, were in control at this moment. Nothing could be more untrue. This council, which represented all the Sioux of Pine Ridge and was every bit as democratic as our own Congress, included all the leading chiefs, the progressives being in the majority. Even American Horse, who was in disgrace with the people for leading in the signing of the hated land agreement, was present and was freely expressing his views in his usual style of oratory. At the moment he was posing as President Harrison's best friend. After full deliberation this council decided to send a party of Pine Ridge men to the Shoshoni reservation in Wyoming, to search out the exact truth concerning the messiah.

The leaders of the Pine Ridge party were Good Thunder and Cloud Horse. They left the agency late in 1889 and returned soon after New Year, 1890, with full confirmation from the Shoshonis as to the existence of the messiah and his wonderful powers. The chiefs counciled and sent out a new party to cross the Rockies in winter and go into the Nevada desert in quest of the messiah. We may put it thus definitely, but these Indians knew nothing of the lands beyond the Shoshonis, having only a vague and fantastic idea of immense snowy mountains and a lonesome desert land inhabited by Indians the Sioux knew only by rumor. The second party, like the first, was made up of rather prominent men: Good Thunder, Cloud Horse, and Yellow Knife of Pine Ridge, Short Bull, a minor medicine man from Lips' camp on Pass Creek at Rosebud, Flat Iron, Yellow Breast, and Broken Arm. On their way west these men were joined by Kicking Bear, a Sitting Bull hostile from Cheyenne River. They were also joined by some Northern Cheyennes and other Indians who were seeking the messiah.

There never was a stranger journeying than this of the Sioux and other Plains Indians in quest of a mysterious messiah in lands so remote that they had only a vague idea where they were going. They rode their ponies westward through empty plains toward the Shoshoni reservation, beyond which lay the great mountains which no Sioux had ever penetrated. The Shoshonis were their former bitter enemies; but now all the tribes were brothers in a quest for the Indian Christ who was to save them from the whites. Among the friendly Shoshonis the Sioux and their companions obtained some information about the way they were to take. Going on, they came to a railroad and sat on their ponies in wonder as a train came up. Some cowboys riding in the caboose invited the Indians to come with them; the ponies were loaded on a boxcar and the bewildered Sioux were drawn swiftly onward into an unknown world.

At remote Walker Lake in the heart of the Nevada desert the Sioux came face to face with the messiah. He belonged to a tribe unknown to them, the Paiutes, and he called himself Wovoka, although the whites knew him as Jack Wilson. His father had been a medicine man and prophet; he had added to his father's store of mystery and magic some strange gleanings from Christianity, evidently including features taken from the Shakers and Mormons; now he was the messiah, preaching new hope for his race, which was to be saved from slavery to the whites by the return of myriad hosts of dead Indians. The spirits or ghosts were to come home in the spring of 1891 when the grass was an inch high, bringing with them all the herds of buffalo and other game the whites had destroyed. While the Sioux were with him, early in March, 1890, Wovoka held a great council for all the Indians who had come across the Rockies to find him. He taught them the beliefs and rituals of his new faith and sent them home to spread his teachings among their people.[6]

The Pine Ridge party returned from their visit to the messiah in late March or early April, most of the men firmly convinced that the man they had found was a true messiah. A council was now called to decide what was to be done; but at this moment William Selwyn, the mixed blood who had read the first letters concerning the messiah to the Pine Ridge Indians, became alarmed and warned Agent Gallagher as to what was going on. Gallagher

[6] *Report of the Secretary of War* (1891), I, 191; James Mooney, "The Ghost-Dance Religion," in B. A. E., *Fourteenth Annual Report, 1892–93* (Washington, 1896), 816–21.

promptly arrested Good Thunder and two others. He kept them in jail; but they would not talk. The agent's action, however, caused the Sioux to give up the plan for a general council. Then Kicking Bear, the Cheyenne River man who had been with the party that visited the messiah, came to Pine Ridge with news that the Northern Arapahoes on the Shoshoni reservation in Wyoming had taken up the rites taught by the messiah; they were holding ghost dances and some of the dancers fell dead, went to the spirit land, talked to their dead relatives, and then came to life and told the people where they had been and what they had seen and heard. These Arapahoes were utterly convinced of the truth of the messiah's teaching. The ghosts in the spirit land were preparing to return to join their relatives on earth and to help them against the whites. On hearing this report, the Pine Ridge Indians brushed aside Agent Gallagher's orders and held a council within a few miles of the agency. Red Cloud, forgetting that he was a Roman Catholic now stated that he believed in the messiah and that the people should begin the ghost dance. Other chiefs were of the same opinion, although most of them were Christian converts.

In November, 1890, General T. H. Ruger, commanding the troops on the Missouri River, reported that the ghost dance craze started at Pine Ridge and was spread by Pine Ridge emissaries to the other agencies. This implied that a kind of conspiracy was hatched at Pine Ridge; but that clearly was not the case. The Pine Ridge Sioux were nearest to the Shoshoni reservation in Wyoming, and they had been exchanging visits with the Shoshonis for several years. It was natural that they should be the first Sioux to learn of the appearance of the messiah and the first to send a party out to seek him. When this party returned in the spring of 1890, Agent Gallagher put a stop to the plans for ghost dancing at Pine Ridge, and Kicking Bear and Short Bull then left the agency and went home. These men were both former members of Crazy Horse's wild camp of the 1876 period. They were now fanatical believers in the messiah and were dangerous men, but they were not Pine Ridge Indians. Kicking Bear went home and started the ghost dance in the camps of the former hostiles far up Cheyenne River, over sixty miles from the agency and beyond the control of the agent and his police. Meantime Short Bull went home to Rosebud with his lieutenant, Mash-the-Kettle. They started preaching to the Indians and are said to have held the first ghost dances on the Sioux reservation, at Iron Creek on Little White River about eight miles west of Rosebud Agency. Curiously, Agent Wright did not seem to understand what was going on. In his report of August 26, 1890, he does refer to his efforts to break up old Indian customs and particularly mentions "ghost lodges." His Indians, he states, put up a lodge for a person who has died, a spirit or ghost lodge, and then the people come and give cattle, wagons, and every kind of property, and all the gifts are collected and a great feast for the dead is held and the collected articles are then given away to needy or worthy families. Wright was here describing the ancient Sioux custom of holding a memorial for the dead, and it would seem that he was so badly informed that he thought his Indians were engaged in this quite harmless ceremony when in fact they were falling victims to the new ghost dance madness.

On receiving vague reports of the activities of the ghost dance prophets Short Bull and Mash-the-Kettle, Agent Wright sent his police to the camp on Little White River and had the men brought to his office. Even when he was face to face with the fanatical preachers of the new faith he evidently did not realize that the situation was serious. He ordered them to cease their activities, and they submitted in apparent meekness to his will.

Wright returned to his routine work, to the important task of hustling his Sioux, urging them to plant more acres of crops than ever before. He was happily planning to hold the first agricultural fair ever seen among the Sioux (if drought and hot July winds spared any specimens of corn and vegetables to be exhibited), and he was too preoccupied with this scheme to hear the grass whispering of a strange new Indian Christ far away to the west, in the lands beyond the Yellow Faces.

These Sioux agents of 1890 were amazing persons. With the ghost dance mania sweeping their Indians off their feet and the madness spreading like fire in dry grass, the agents did not seem to know that anything unusual was going on. Farming programs and other routine work were their sole interests. It was the same at the Indian Office in Washington until June, 1890, when the meddlesome War Department disturbed the quiet of that dignified establishment by calling attention to the excited condition of the Crows and certain other tribes, on account of the preaching of an Indian messiah somewhere in the lands west of the Rockies. Charles E. Hyde, a citizen of the town of Pierre across the Missouri River from Cheyenne River Agency, had written a letter warning the Indian officials that a plot was brewing for an uprising among the Sioux. The officials now dug up this letter and sent circular instructions to all the Sioux agents for a report on the subject. The reply indicated that the agent at Rosebud knew nothing of any impending trouble; the agent at Cheyenne River stated that there was some excitement among his Indians concerning a messiah but that he was not much concerned; Agent Gallagher of Pine Ridge reported the messiah craze as nothing serious. McLaughlin of Standing Rock reported some excitement at the other agencies, but asserted that his own Sioux were not affected. He did not believe that the Sioux were planning any disturbance. In this report, June 18, 1890, McLaughlin suggested that it might be well to remove certain nonprogressive leaders from the reservation. He wished to remove Sitting Bull, Circle Bear, Blackbird, and Circling Hawk from his own agency; Spotted Elk (alias Big Foot) and his lieutenant from Cheyenne River; and from Rosebud, Crow Dog, the murderer of Spotted Tail, and Low Dog, a former hostile who had returned home with Big Road about 1882, when the Sitting Bull hostiles surrendered. In effect, McLaughlin (ranking as the Sioux agent longest in service and termed by his admirers the best agent the Sioux ever had) seemed to think the messiah craze was a small matter, but he was inclined to make it the excuse for banishing some of the leading enemies of progress from the reservation. The Indian Office, however, could not see its way to removing any of the Sioux leaders from their homes simply because an agent had labeled them nonprogressive. McLaughlin in his list had not included the original ghost dance spreaders, nor had he blacklisted any of the reservation chiefs. The two chiefs on his list were Sitting Bull hostiles. He reported that he had consulted with his most trustworthy Sioux and that they had all expressed strong disbelief that there was a secret plan for an uprising. They said that if there was such a plan, they would be sure to know it; for the Sioux could not keep a secret, and that was true.[7]

[7] *Report of the Commissioner of Indian Affairs* (1890), 328.

10
Whirlwind

NO INGENUITY could have invented a more perfect means than this strange new faith to lift the Sioux out of the deep depression and hopelessness about the future that the events of 1889 had cast over their minds and spirits. At Rosebud, when Short Bull returned from his visit to the messiah (March–April, 1890) and began to preach, the drooping Sioux lifted their heads and people flocked to listen to the new prophet. Most of the Sioux who heard believed his message, and they eagerly became his pupils, learning rites and songs and the methods for conducting the ghost or spirit dances. The dances were hypnotic in their effect. Men, women, and children formed a great circle, all holding hands and revolving quickly toward the left, chanting the sacred songs as they went. At every performance some dancers fell in a swoon, and the medicine men said that they were dead. When they were revived they stood quivering with emotion, telling of their journey to the spirit land, their meeting with dead relatives and friends, and their talks with the dead. At these ghost dance gatherings the dancers fasted for many hours in preparation for the ceremonies; they went through the weakening process of the sweat-lodge purification. The dancing continued on four consecutive nights; on the fifth night they danced until dawn, and it was then that most of the dancers swooned and were pronounced dead. After such a gathering everyone went home; but they reassembled at the end of six weeks to go through the same program again.

The whole of this procedure was so typically Indian in thought and method that it took an amazing hold on the Rosebud Sioux.

They had no doubt at all that the dancers who swooned actually went to the spirit world and visited their dead relatives and friends. The hysterical statements made by these dancers when they returned to consciousness convinced even the most levelheaded Indians that what they said was the simple truth. These Indian men and women gave minute details of the personal appearance of long-dead relatives and friends, described their clothing and personal belongings such as weapons, tobacco bags, and pipes. Many of the listening Sioux who had known the dead persons knew that all the little details were correct. And all the dancers who had gone to the spirit world brought back the same message. The spirits of all the dead were returning to earth, to join their living relatives and assist them in a return to the old happy life of hunting and wandering freely in a land from which all whites had disappeared.

As the Rosebud Sioux listened to the revelations of their friends who had visited the spirit land, hope flowed into them like a swelling tide. But they were not yet confident enough in their new strength to flout authority, and when Agent Wright sent his Indian police to break up the ghost dance meetings, the dancers took fright and fled. The prophet, Short Bull, was taken by the police to the agent's office, where he was reported to have meekly promised to cease his preaching. He merely went underground, retiring to the more distant Sioux camps where the police could not reach him. Agent Wright soon forgot the ghost dance matter and returned to his original interest of urging his Sioux to increase their farming efforts.

At the other agencies the course of events followed the pattern described at Rosebud. At Pine Ridge, Agent Gallagher easily stopped the first ghost dances in the camps near the agency; but presently a new ghost dance prophet named Porcupine came home from visiting the messiah and started dances on Wounded Knee Creek, where the former Crazy Horse hostiles, now led by Big Road, were among the first to take up dancing. Here, as at the other agencies, it was generally the former hostiles of 1876 who gave the ghost dance mania its first impetus. Those Sioux who had been on the reservation before 1876 still retained enough respect for authority to refrain from flouting the agents' orders forbidding ghost dancing.

The Sioux were still suffering severely from want of food. They waited for their crops to mature, hoping to add to the slender rations they were receiving; but in midsummer drought began to destroy the crops, and then the Indian Office ordered a new cut in the beef issue. Lack of funds and the failure of Congress to pass the Sioux appropriations on time were the excuses for this cruel action. The effect on the Sioux was apparent at once. At Rosebud, where the Indians had quieted down after Agent Wright stopped the ghost dance in May, the Sioux now began to denounce their white oppressors. Even the so-called progressives raised their voices in anger. Was this the time, with everything going wrong, for a fool in the Great Father's town to cut the beef ration a second time in less than a year? The Sioux at Rosebud were getting out of hand, and now an incident occurred that gave clear indication of the direction of the drift. The Indians were all assembled at the agency on ration day, when a report was brought in that white soldiers had invaded the reservation. Before Agent Wright could take any action, all his Sioux rushed to arms, mounted their ponies and rode furiously away, to meet and turn back the soldiers. Wright gathered his Indian police and pursued the warriors. When he finally overtook them, he found that they had stripped themselves naked, as for battle, and were riding up and down, furiously waving Winchester rifles as they watched for the approach of the white soldiers. Wright had a hard time to convince the Sioux that the report of troops coming was false.[1]

As to what happened next at Rosebud there is considerable uncertainty. The missionary, the Reverend W. J. Cleveland, stated that the agent was relieved, which can only mean that he was removed to make way for a new agent. Yet Wright returned a year later and resumed his position as agent, as if nothing had happened. Then, in 1891, Wright reported that in the preceding summer, 1890, "complications arising necessitated the agent to leave the agency for a time, and an acting agent was placed in temporary

[1] *Report of the Commissioner of Indian Affairs* (1891), 411; *Report of the Secretary of War* (1891) I, 140.

control." Temporary was not exactly the right word. Wright was gone for a full year, a year of one violent crisis after another which made his presence imperative; then he returned with no explanation beyond some carefully trimmed phrases. Was he the first victim of Senator Pettigrew's slaughter of agents, and did Pettigrew put him back in 1891 after having his knuckles soundly rapped?

The Harrison administration that came into office in 1889 was pledged to civil service reform, but it did not keep its pledges. For the Indian Service it invented a new and noble policy which it termed home rule; and as far as the Sioux were concerned, this meant simply that the South Dakota reservations were turned over to Senator R. F. Pettigrew, who at once began the pleasant task of weeding out Democrats from New York, Ohio, and other states, replacing them with South Dakota Republicans. Starting with small people, agency clerks, farmers, and teachers, the Senator happily worked his way up toward the agents themselves, and then he plucked them, one after the other, leaving only McLaughlin—the Standing Rock perennial whom neither political party dared to touch. With incredible disregard for public interest and even for safety, Pettigrew removed all of the experienced Sioux agents in the midst of the ghost dance excitement, replacing them with inexperienced South Dakota men. His removal of Agent Gallagher of Pine Ridge was his crowning achievement.

D. H. Gallagher was an Irishman from Indiana who had won the rank of colonel in the Civil War. When the Democrats removed Agent McGillycuddy of Pine Ridge in 1886, they put in an army officer as acting agent and then quietly introduced Gallagher as the new agent. He did not have the mint mark of approval placed on him by the leaders of the Indian welfare groups, who could not forgive him for succeeding their admirable McGillycuddy. They watched him like cats after a mouse; but they had to have an excuse for assaulting him, and they could allege nothing against Gallagher further than he was easy going and had relaxed the iron discipline that McGillycuddy had maintained during seven years. Gallagher, indeed, had given the Pine Ridge Sioux what they most needed: some quiet and an end to the hustling and shouting they had endured in McGillycuddy's day. Red Cloud in 1890 was the same stubborn Sioux who had stood toe to toe with McGillycuddy, exchanging resounding blows, but he was no longer the chronic troublemaker. He and the Irish colonel could disagree, as they did over the land agreement of 1889, without causing the heavens to rock and the troops at Camp Robinson to stand to arms. The aging chief was now a Christian, after being kept a heathen for many years simply because McGillycuddy had outlawed the Roman Catholic religion at Pine Ridge. Where McGillycuddy had given his Sioux the choice of becoming Episcopalians or remaining heathens, Gallagher had established religious freedom. He was himself a Roman Catholic, and naturally some narrow-minded folk blamed him for letting down the bars to that faith. McGillycuddy had simply told inquiring priests that it was an Episcopal agency and that if they attempted to start a mission, he would have his armed police remove them from the reservation.

Red Cloud's Catholicism did not prevent his listening to the call of the Indian messiah in the spring of 1890, and presently he announced that he believed in the messiah's message. He told the Pine Ridge Indians that if they wished to start ghost dancing, they were free to do so, but Gallagher said "no" and sent his police to enforce his order. If the trouble caused by the ghost dance at this agency did not end there, we can hardly blame the agent, or even Red Cloud. The agent did what was possible to control his Indians, and Red Cloud soon recanted his pronouncement in favor of the messiah. Curiously, when the craze flared up again in summer, it was Little Wound, a notable progressive and supposedly a loyal Episcopalian, who led the Pine Ridge ghost dancers. Now nothing could stop the mania, and it was the same at the other big Sioux agencies. Those who have been inclined to accuse Gallagher of Pine Ridge of letting things down from the splendid condition of discipline McGillycuddy had maintained do not take into account that in the summer of 1890 conditions at all the agencies were much the same. If the Pine Ridge police under Gallagher were inefficient (which they were not), the same condition among the police of the other agencies must be admitted, and that cannot be done without a complete disregard for the facts. Facing an impossible situation in 1890, most of the Sioux police were ready to risk their lives to support white officials who paid them the pitiful sum of ten dollars a month. The crux of the matter lay in the fact that the Sioux had never approved the Indian police system. They put up with it in quiet times; but in 1890 when most of the Indians seemed to have gone mad their hatred for the police flared fiercely, and no group of Indian police could then approach a Sioux camp without the risk of bringing on a fight with the odds ten to one against them.

When the real trouble started in the summer, Agent Gallagher had the worst situation that any Sioux agent was faced with. Sheer bad luck brought the Sioux fanatic named Porcupine to his agency, to start new ghost dancing in camps on Wounded Knee, twenty-five miles northeast of the agency. Big Road, who had led the Oglalas in the Crook and Custer battles in 1876 and had later fled to join Sitting Bull in Canada, was now living with his former hostiles here on Wounded Knee, and it was under his leadership that the ghost dancing was started. The Sioux here danced at irregular intervals, the excitement becoming more and more intense. While engaged in these ghost dances, many of the Sioux "died," as they put it, and when they came to themselves again, they were frantic with joy. They had been in the spirit land; they had talked to their dead relatives and friends, who were alive and happy and were preparing to return to earth, bringing vast herds of buffalo and other game with them. These Pine Ridge Indians were suffering from hunger and great depression of mind, and this promise of sure help about to come from their ghost relatives lifted them into ecstacy. The mania spread like flames driven by a great wind. Soon the center of excitement shifted westward from Wounded Knee, and No Water's nonprogressive camp near the mouth of Big White Clay Creek, twenty miles north of the agency, became the Mecca of the ghost dancers. Jack Red Cloud, the old chief's son, was a leader here, where in mid-June ghost shirts are said to have been first worn by the Sioux. Made of animal skin or cloth painted with sacred Indian symbols, these shirts were supposed to protect the wearers from all harm coming from the whites. Bullets could not penetrate them, and the maddened Sioux no longer feared the agent's police or even the white soldiers. It was into this welter of madness at Pine Ridge that the complacent Commissioner Morgan now cast the affront of a new cut in the beef ration. Drought was already destroying the Sioux crops, and many progressive Indians who were hesitant about joining the dancers were now faced with the choice of doing that or of sitting down to watch their families starve. There seemed to be no hope of aid from the whites. Congress, otherwise engaged, had not even passed the Sioux appropriation bills, and the agents were scraping the bottoms of the barrels, trying to get along on almost nothing until money for fresh supplies was voted.

In late August the ghost dance excitement in the No Water camp, some eight miles north of Pine Ridge near the mouth of Big White Clay Creek, was so intense that Agent Gallagher knew that he must take action; but before acting he called a conference of all the white men at the agency. These men agreed with the agent's view, that he should go to the No Water camp with a force of police and attempt to break up the ghost dance. Philip F. Wells, the agency interpreter, was the only man who disagreed with this view. He was also the only man present who had ever seen the Sioux stripped for fighting and in a desperate mood. He told the agent that if he took his police to this camp there would be a massacre. His own advice was for the agent to go to the camp with Red Cloud and Man-Afraid-of-His-Horse, the two leading chiefs, with Wells himself as interpreter, all without arms, and to try to come to a friendly agreement with the ghost dance leaders. Gallagher said bluntly that there was not a trace of friendliness or agreeableness in any of the ghost leaders, and that he was going to that camp with his police.

He set out on Sunday, August 24, taking Philip Wells and thirty armed police. On the way they met a Sioux who warned them that the No Water camp was swarming with ghost dancers, all spoiling for a fight; but when they reached the group of scattered log cabins that was the center of the No Water settlement, not a soul was in sight except a few frightened Indian women and children. Then Wells spotted an old Sioux man whom he knew, hiding in the creek bed. Only the head and shoulders of the Indian stuck up above the creek bank, but Wells could see the rifle he held gripped in his hands. Wells pointed him out to Gallagher, who issued a sharp order for Lieutenant Fast Horse to go with some of his men and arrest the Indian for daring to flourish a gun in his agent's face. Gallagher could not even give an order to his police without having Wells turn his words into Sioux. Wells translated the order for Fast Horse, then added: "Don't you do it! The agent doesn't know that the creek bed is full of Sioux and that we are in a trap."

Fast Horse hesitated. He was sworn to obey the agent, but he did not like Gallagher's order. "Do as I tell you!" snapped Wells. "You have all these policemen for witnesses that I'm taking the responsibility." He then urged Gallagher to let him go out and get hold of this old Sioux man without any gunplay. The agent gave a grudging consent, and Wells laid down his rifle and started walking slowly toward the creek. He called the old man "uncle" and addressed him with careful respect. It might have been funny if the situation had not been as tight as a drumhead. Wells would have the old Indian ready to lay down his gun and come out for a folksy talk, and then Gallagher (who as a former colonel in the army could not tolerate seeing a subordinate taking the lead) would stride out and get in front of Wells. "Uncle" would promptly grip his rifle and assume a hostile pose. In the end, Wells induced the colonel to stand back, and the old man dropped his gun and came out.

Then the head and shoulders of one ghost dancer after another rose above the creek bank. There was a surprising number of them, and they were all armed. Wells got some leaders out for a talk; but it was soon apparent that Gallagher's view was correct. There was no possibility of friendly agreement. These ghost dancers were defiant; they intended to go on with their dancing, and they would not budge from that decision. Gallagher thought it over. The ghost dancers outnumbered his police at least ten to one and held a covered position in the creek bed. His men were in a trap and unwilling as he was to do so, he gave the order for the return to the agency.

Philip Wells thought it unfortunate that the agent had not

taken his advice in the first place. He believed that there was only one hope left for peace, and that was to use Red Cloud and Man-Afraid-of-His-Horse and find out whether these Sioux were too maddened by the ghost dance mania to listen to the voice of authority coming from their tribal leaders. If that failed, calling in the military was the only resource left.

Red Cloud's white critics stated later that at this juncture he told the Sioux to go on with the ghost dance. Perhaps he did; but why should he be singled out as a culprit when nearly all of the chiefs were being dragged into this mania by their own followers? Little Wound, a progressive and Red Cloud's enemy, stated later: "I told them they could dance, and if it was a good thing, well, and if it was bad it would fall to the ground." He went much farther than Red Cloud, becoming an active ghost dancer. At this time his daughter was dying, as he claimed of hunger and the failure of the agent to provide any medical care; and watching his child wasting away, the old chief turned his back on the whites and led his band into the ghost dance camp.

It was at this critical moment that Senator Pettigrew of South Dakota, who had charge of political patronage connected with the Sioux agencies, removed Agent Gallagher on the pretense that he was responsible for the breakdown of control at his agency and appointed in his place Dr. D. F. Royer, a Dakota Republican. This was pure political jobbery; for Royer was utterly inexperienced and evidently a man of much less strength of character than Gallagher. The latter at any rate was known to be a brave man; but Royer, as soon as he reached the agency, let the Sioux see that he was afraid of them, and within a week he was writing to the Indian Office concerning the hopelessness of the situation. He asked for troops. The other agents did not think troops were needed, although conditions at Rosebud were as bad as at Pine Ridge, while at Cheyenne River the fanatic Kicking Bear had most of the former hostiles wild with excitement and engaging openly in ghost dancing. At Standing Rock that most experienced of agents, James McLaughlin, reported that all was quiet; yet only a few days later (August 17) he had to report that the ghost dance, moving northward from Cheyenne River, was infecting his Indians.

Sitting Bull, like McLaughlin, ignored the messiah craze at first. All through the spring and early summer of 1890 he was busy trying to keep aflame the Sioux discontent over the land agreement of 1889, and when he heard the first rumors of the messiah, he seems to have discounted the whole thing. But by August the mania had spread among the Sioux to such an amazing extent that he was impressed; and he now contacted the ghost dance prophet, Kicking Bear, at Cheyenne River.

In October, McLaughlin was shocked to learn that Kicking Bear was conducting ghost dances in Sitting Bull's camp on Grand River, forty miles southwest of Standing Rock Agency. The agent now sent half of his police force, thirteen men headed by a captain and second lieutenant, to arrest Kicking Bear as an unauthorized intruder on his reservation. The police went to Sitting Bull's camp keyed up to do or die; but on October 14 they returned to the agency without their prisoner and looking like men who had experienced a violent shock and who were still in a dazed condition. At first they would not talk; but after a while the captain and lieutenant recovered sufficiently to explain to the agent what had happened. They had gone to Sitting Bull's camp and had found the messiah faith being openly preached by Kicking Bear. They had watched the ghost dance and had seen Sioux fall dead in the circle of dancers, had seen them come to life again, and had listened to their ecstatic accounts of their visit to the spirit land. These Sioux had described in minute detail the appearance of their relatives and friends in the spirit land, and the amazed police, who had known most of these Sioux when they were alive, had recognized them and had even remembered some of the articles of clothing, pipes, and weapons that were now described. As they watched and listened, the Sioux souls of these policemen stirred and awoke and their faith in the whites and their teachings was shattered for the moment. They left the camp without attempting to execute their orders, and even after a forty-mile ride back to the agency they were still in a kind of trance.

Perfectly assured of the loyalty of these policemen, McLaughlin was deeply shocked that anything the Sioux might say or do could cast them into such a condition of mental helplessness. He thought it over carefully and then selected Lieutenant Chatka, a very determined man, and sent him to Sitting Bull's camp with just one companion, with orders to see to it that Kicking Bear and his six Cheyenne River ghost dance leaders should leave at once. Chatka and his helper either did not have Sioux souls or else they had ones of sufficient toughness to resist the pull of the ghost dance mania. They grimly faced the mass of excited Sioux in Sitting Bull's camp and spoke their orders; Kicking Bear and his friends, seeing that these two policemen meant to be obeyed, forgot their holy mission and meekly submitted. But that did not end the matter, for three days later the agent learned that Sitting Bull was proclaiming himself the head of the ghost dancers of Standing Rock Agency and that he had told a large gathering of Sioux that he was ready to fight and die for the new faith. He was now holding continuous ghost dances at his camp, where the Sioux in a frenzy of religious excitement were behaving like drunken persons, their conduct as McLaughlin officially reported being amazing, silly, and disgusting. Small children taken from school were being led by their parents into this maelstrom of madness. Through a nephew of Sitting Bull's named One Bull, McLaughlin sent a message to the old prophet, inviting him to the agency for a talk; but Sitting Bull, probably fearing arrest if he left his own camp, refused to come.[2] He remained in his distant camp, surrounded by armed fanatics. The ghost dance spread swiftly from end to end of Standing Rock reservation; the schools had to be closed, and only in the camps of Yanktonais were white men safe. With Fort Yates situated on the reservation, immediately south of the agency, McLaughlin and his employees were in no danger. There were four companies of the Twelfth Infantry and two troops of the Eighth Cavalry at this post.

McLaughlin on October 17 wrote a full report of the dangerous condition of affairs and sent it to the Indian Office. Commissioner Morgan was in the field, visiting the southern Indian reservations, busy with plans for improving the Indian schools, and McLaugh-

[2] McLaughlin's reports in *Report of the Commissioner of Indian Affairs* (1891), 329. These reports tell a somewhat different story from that in McLaughlin's book, published some years later.

lin waited twelve critical days before he received a reply from Acting Commissioner Belt—and such a reply! Belt had submitted McLaughlin's report to the Secretary of the Interior, and that lord of the cabinet now sent instructions that McLaughlin was to inform Sitting Bull that he would be held personally responsible for any trouble at Standing Rock. Would these officials ever learn that they were dealing with Indians, and not with white men who understood and feared the law?

McLaughlin was in no haste to take this ladylike admonition to Sitting Bull's camp, where the frenzied Sioux were all armed and not in the least afraid to die. Had they not their prophet's assurance that if they were killed now they would return in triumph with the vast army of Sioux ghosts in the spring of 1891? On November 18, McLaughlin at last visited Sitting Bull's camp and beheld with his own eyes the amazing spectacle of the ghost dance. About one hundred Sioux, including boys and girls taken from school, were holding hands in a great circle and going around, jumping toward the left, revolving about a sacred pole or tree. Sitting Bull and his religious helper, Ghost Bull, were seated in the doorway of a sacred tipi watching the wild performance. McLaughlin, who had lived among the Sioux most of his life, was almost as dazed at this spectacle as his Indian police had been. This dance was not slow like all the old Sioux dances; it was fast and violent, the dancers leaping to the left, their muscles jerking, their breath coming in panting gasps, their tongues lolling out. As they whirled they chanted. A school girl fell in a faint: "dead," they said. Two ghost dance officials carried her to the sacred tipi, and presently she came to life and rushing out started a frantic harangue, telling the frenzied Sioux of her visit to her friends in the spirit land. McLaughlin, watching this incredible scene, decided not to give Sitting Bull the Secretary of the Interior's message about holding him personally responsible.

Perian P. Palmer, Senator Pettigrew's new appointee as agent at Cheyenne River, took charge in August and was not at first worried over the ghost dance situation. His agency, like McLaughlin's which bordered it on the north, had a military post within sight of the office windows: Fort Bennett, a small post, while across the Missouri and a few miles below the agency was Fort Sully, a larger garrison. The white men at the agency felt safe and did not regard the ghost dance excitement as a serious danger; but they wished to see Hump brought in and put under military constraint. This chief, the friend of Crazy Horse, had taken a leading part in the war of 1876; he was the big man among the former hostiles now settled on upper Cheyenne River, and the white men at the agency looked on him as a most dangerous character. Their view seemed to be correct. Kicking Bear, the ghost dance prophet, submitted when Agent Palmer ordered his police to arrest him. He was tried by the Indian court, but his Sioux judges released him. Still, the whites at the agency did not fear his influence greatly. They believed in late summer that the ghost dance mania was beginning to die down. All the Sioux close to the agency, along the Missouri (mostly Two Kettles and Sans Arcs), were progressives and perfectly friendly. It was among the wild Miniconjous of upper Cheyenne River that Kicking Bear had influence. In September he had the ghost dances going at full swing in the camps of Hump and of Big Foot (son of old Lone Horn, the Miniconjou head chief who had died in 1874).[3] This Sioux prophet had abandoned the peaceful teaching of the messiah and was telling the maddened Sioux that ghost shirts were bullet proof and that they now had the power to destroy the whites whose troops could not harm the Indians. Here at Cheyenne River as at the other agencies the one thing that shocked the whites most was the collapse of all authority. The agents whose word had been law among the Sioux were now flouted and the armed Indian police were in general unable to execute their orders. At this agency the police did not dare to go to Hump's camp near the mouth of Cherry Creek or to Big Foot's camp farther west, at the forks of Cheyenne River. Another feature of this situation lay in the fact that the Sioux had turned once more to their chiefs for leadership. All the carefully laid plan of the vision-

[3] Old Lone Horn is said to have died in his camp on upper Cheyenne River near the Black Hills in the winter of 1874–75. His sons, Touching-the-Clouds and Big Foot (alias Spotted Elk), were with the hostiles during the war of 1876, Harry Anderson thinks that these two chiefs, Touching-the-Clouds and Big Foot, were probably half-brothers, sons of Lone Horn by different mothers.

ary Indian welfare leaders to destroy the chiefs and put all authority in the hands of white agents, all the labor of the Washington officials and the agents to carry out this plan now turned to dust. Even the Sioux progressives put themselves back under the leadership of their chiefs and would obey no one else. Almost in a day the planning and work that had been carried on since 1878 with the object of ending the Sioux tribal organization was destroyed by this messiah craze, and the Sioux slipped back to the condition they had been in before the reservation was formed.

Commissioner Morgan, whose presence in Washington was imperatively needed, was still visiting the southern reservations, immersed in his passion for tinkering with the Indian school system, while the Secretary of the Interior, after sending his absurd message to Sitting Bull, took no further action. For a time the Sioux agents imagined that the ghost dance mania was dying down. They sent more encouraging reports to Washington; but in the second week of November a veritable storm of telegrams reached the Indian Office, reporting a complete breakdown of control, with large gatherings of maddened ghost dancers defying the agents and ready to resist with arms if interfered with. The Washington officials were aghast. The mere thought of having to call in the troops was death in their hearts, an open confession that all their boasts of progress among the Sioux were hollow. To the leaders of the Indian welfare groups the use of troops against Indians was the supreme shame, the last bitter confession of failure; yet by November 10 the officials in Washington were seriously considering that step. Major General Nelson A. Miles, in command of the army, had already quietly sent officers to each of the Sioux agencies, to report on conditions.

The main danger lay at Pine Ridge and Rosebud. The new agent at Pine Ridge, Dr. Royer, was displaying a woeful lack of competence. Almost from the day he took charge he had been urging the calling in of troops. The Indian Office would not consider such action, but Royer was at length advised that General Miles was on his way West on routine army business and would stop at Pine Ridge for a conference. The General told Royer that he did not consider the danger serious, but Miles should have been warned by the attitude of the Pine Ridge chiefs. When he held a council with them and urged them to stop the ghost dancing, their only reply was that the Sioux intended to go on dancing. Miles gave them a severe lecture, telling them that the dancing must stop; but they made no definite promises.

At this critical moment Young-Man-Afraid-of-His-Horse, the leading progressive chief, left Pine Ridge to go on an extended hunt in Wyoming. He either thought, as Miles did, that the situation was not dangerous, or else he was washing his hands of the whole affair. On October 20, Agent Royer sent another excited telegram to the Indian Office, demanding troops and stating that six to seven hundred soldiers were needed to control his Indians. No action was taken; on November 11, Royer telegraphed again, asking permission to come to Washington to explain the situation personally to the officials. Permission was refused; but day after day the alarmed agent pelted the Indian Office with telegrams, urging that he should be permitted to come East for a conference.[4]

In strange contrast with Royer, Acting Agent E. B. Reynolds at Rosebud was very quiet and unperturbed; yet he had every bit as much cause to shout and rush about as his excited neighbor at Pine Ridge. Short Bull, the Rosebud ghost dance leader, had been warned in early summer to stop his preaching, and he had apparently obeyed. But then a second cut in the beef ration was made, and after that drought began to destroy the Rosebud crops. Short Bull took instant advantage of the fresh outbreak of discontent among the hungry Indians. He went to Black Pipe Creek and started ghost dancing. Instead of shouting with alarm like Royer of Pine Ridge, Agent Reynolds drew a curtain of silence around his reservation; even today it is difficult to follow the course of events. All that is known is that the outbreak of ghost dancing on Black Pipe Creek came in September and that Reynolds sent a policeman—just one—to Red Leaf's Wazhazha camp on Black Pipe, to stop the ghost dance. The policeman got home alive, reporting no results, and the annoyed agent then sent ten policemen to Red Leaf's camp. They found the place alive with Sioux, maddened by the ghost dancing, the men armed with Winchesters

[4] Mooney, *The Ghost-Dance Religion*, 848.

Squaw Dance at the Rosebud Reservation, 1889. Reprinted, by permission, from Hamilton and Hamilton, The Sioux of the Rosebud, *pl. 111.*

Indians on Horseback Awaiting the Release of Government-issue Cattle, to Be Shot and Butchered on the Prairie, About 1889. Reprinted, by permission, from Hamilton and Hamilton, The Sioux of the Rosebud, *p.* 64.

and their naked bodies festooned with belts of brass cartridges. Having enough sense not to start a fight, the ten policemen also got home alive.

Reynolds now wrote to the Indian Office, stating that for weeks past his Indians had been spending every dollar they could get hold of for brass cartridges and that they had also traded many ponies to white men for ammunition. This had a very ugly look. Moreover, Two Strike, the head of Brûlé Band Number Two and the strongest chief at Rosebud since old Spotted Tail's murder, had brought his band together (they had been scattered out in farming districts by Agent Wright) and was considering joining the ghost dancers. His lieutenant, Yellow Robe, seems to have favored such action, and that perennial troublemaker, Crow Dog, was in the camp, daily urging Two Strike to move to Black Pipe Creek, where the ghost dancers were assembled.

On October 31, Short Bull harangued the ghost dancers in Red Leaf's camp, and it now became apparent that from being a follower of the Nevada messiah this man had lifted himself to the rank of messiah in his own right. He told the crazed Sioux that, since Agent Reynolds was trying to stop the dancing, he was going to punish the whites by advancing the date of their destruction. The Nevada messiah had said that in the spring of 1891 a great dust storm would bury all the whites; then new grass would come and all the ghosts of the Indians and game animals would return to earth. Short Bull now stated that he was advancing the time of the dust storm to the new moon after the next new moon—about December 11, 1890. He assured his listeners that if they kept faith and continued dancing the whites could not harm them. Bullets could not penetrate their sacred ghost shirts. They must desert their homes and all encamp on Pass Creek, the boundary between Rosebud and Pine Ridge. At the mouth of the creek a tree (a sacred symbol of the messiah faith) was sprouting, and around this tree they would encamp, to await the coming of their ghost relatives. They must not bring any earthly goods or anything coming from the whites with them. All that was to be destroyed.[5]

With Short Bull thus preparing to march the Rosebud ghost

[5] *Report of the Commissioner of Indian Affairs* (1891), 128.

dancers to Pass Creek on the Pine Ridge border, the mania at the latter agency was spreading eastward from the camp of Big Road's former hostiles on Wounded Knee and had now infected the Sioux of Little Wound's band on Yellow Medicine Root Creek, not far west of Pass Creek. Thus it seemed that the Rosebud and Pine Ridge ghost dancers would soon be in touch and the situation would become grave indeed. That old Little Wound should have been led into this craze was really amazing. He and his band were progressive followers of the white man's ways and most of them were rated as Christians. They had a fine settlement on Yellow Medicine Root, forty miles northeast of Pine Ridge Agency, with many small farms, a school, and the handsome little St. Barnabas Episcopal chapel, built and equipped with funds supplied by the estate of Mrs. John Jacob Astor. Yet the messiah craze had infected this model community; Little Wound had joined the ghost dancers himself, had fallen "dead," had visited his relatives in the spirit land, and had now come back to life to speak and act like a wild Sioux of the olden time. With as much spirit as the former hostile Big Road, the old chief defied the agent to interfere with the ghost dancers.

This situation so alarmed Agent Royer that early in November he issued an order for all the white and mixed-blood employees in outlying camps of the Pine Ridge reservation to abandon their school, farm, and other activities and come to the agency with their families. Most of the white employees were so frightened that on reaching the agency they kept on going and did not stop until they had their families safe in the small towns along the railroad in northern Nebraska. Royer had also ordered the friendly Indians to leave their little farms and concentrate in a camp near the agency on Big White Clay Creek. This was called the friendly camp, although many of the Sioux in it were openly siding with the ghost dancers and acting as spies who reported on the activities of the agent and his police.

On November 12, Royer telegraphed the Indian Office that on that day two hundred maddened ghost dancers had seized control of his agency and that he must have troops at once. His message was submitted to President Harrison, who made the astonishing comment that the army was investigating conditions on the Sioux reservation and that for the present the agents should limit their activities to separating the friendly Indians from the ghost dancers, at the same time avoiding any action that might cause irritation. In effect, they were to continue the policy of the past months, a policy of drift. Three days later, on November Royer telegraphed:

> Indians dancing in the snow and are wild and crazy. I have fully informed you that employees and government property at this agency are at the mercy of these dancers. Why delay for further investigation? We need protection and we need it now. The leaders should be arrested and confined in some military post until the matter is quieted and this should be done at once.[6]

Again no action was taken in Washington. November 17 was ration day at Pine Ridge, and thousands of Indians crowded the agency. Among them was a ghost dancer named Little[7] who had been particularly troublesome, and when Royer learned of his presence, he ordered the Indian police to arrest him. It was after dark, and the police found Little surrounded by a group of fanatical ghost dancers. With more courage than discretion the Sioux police attempted to seize Little, and in an instant the enraged ghost dancers closed in on them, shouting to kill them and all the whites at the agency. American Horse, a fine-looking chief even in the cheap citizen clothing supplied by the government, pushed into the milling mob and took his stand in front of the embattled policemen, crying out passionately that no Sioux should shed the blood of a brother Sioux. Jack Red Cloud, openly siding with the ghost dancers in an attempt to win a claim to leadership in some degree comparable to the reputation of his great father, rushed up and thrust a revolver in the face of American Horse, shouting that he was the man who had betrayed the people by signing the land agreement of 1889. American Horse, little impressed by a

[6] *Ibid.*

[7] John Colhoff of Pine Ridge (an excellent informant) tells me that Little had another name, *Cetan Wetakpe* (Charging Hawk) and that he died at the village of Oglala north of Pine Ridge in 1932.

man who talked first and shot afterward, went straight on with his impassioned appeal for peace; and presently a majority of the crowd came over to his side. The ghost dancers, who were standing with the muzzles of their rifles pressed against the policemen's bodies, were induced to stand back, and the police withdrew.[8]

Late this same night Dr. Charles A. Eastman, a fullblood Sioux who had been recently appointed physician at Pine Ridge, was summoned to the agent's office, where he found the agent, a government Indian inspector who had just finished looking into conditions at Pine Ridge, the chief clerk, and a Mr. Cook, a Sioux Indian who was Protestant Episcopal minister at the agency. Royer asked the advice of these men as to the need for calling in troops. The white men all approved of it; but Eastman and Cook gave their opinion that there was no plan among the Sioux for starting a war and that it was inadvisable to call for troops. Royer then said that the calling of troops was approved. He called in Sword, chief of police, Thunder Bear, lieutenant of police, and American Horse, who was staying at the agency because he and his family were not safe in the friendly camp within sight of the agent's office windows. These three Sioux were asked for their opinion, and they all approved of the call for troops. When the Indians left the office, they said to Eastman that they believed the troops at Camp Robinson, Nebraska, had already been ordered to march and that the agent had called his conference only to win support for a move he had already made.[9]

We now come to that incredible incident, the flight of Agent Royer and his dramatic appearance in the little town of Rushville, Nebraska, where he is said to have come roaring down the street in his buckboard with his team at a full gallop and white with lather, shouting as he tore past that the Sioux were up and that everyone would be murdered. The army reports refer to this incident briefly; eyewitnesses have given colorful accounts of Royer's swift passage down the main street of Rushville; but men like Dr. Eastman and James Mooney, who were in govern-

[8] James P. Boyd, *Recent Indian Wars*, Philadelphia, (1891), 224; Mooney, *The Ghost-Dance Religion*, 788; Charles A. Eastman, *From the Deep Woods to Civilization* (Boston, 1916), 83, 93.

[9] Eastman, *From the Deep Woods to Civilization*, 98.

ment service, made no mention of Royer's flight. No date is given; but Royer was at the agency on the night of November 17 and he was back at Pine Ridge with the troops on the nineteenth. His Paul Revere ride through Rushville, if it really took place, must have been on November 18.

On the morning of the nineteenth white men and Indians at Pine Ridge observed a dust cloud billowing up from behind the pine-dotted ridges south of the agency, and within the hour the Negroes of the Ninth Cavalry had taken over control at Pine Ridge. Infantry from Fort Omaha arrived later; within a few days Brigadier General John R. Brooke was at the agency in command of a battalion of the Ninth Cavalry, eight troops of the Seventh Cavalry, eight companies of the Second Infantry, and one battery of the Fifth Artillery. The Sioux in the so-called friendly camps close to the agency looked on sourly as the soldiers took over control. The Indians blamed Agent Royer bitterly for bringing in the troops, and they gave the agent the handsome name of *Lakota-Kokipa-Koshkala:* Young-Man-Afraid-of-His-Sioux. Some of the friendlies fled when the first troops appeared, but there was no panic. Red Cloud was very glum, telling everyone how the soldiers had stolen his horses in 1876. Now in 1890 he was predicting that the soldiers would steal his ponies again, and his band put their four hundred animals inside a large corral and made a great parade of guarding them from the troops. But it was apparent that the sole object of the soldiers was to guard the agency from attack. They threw a picket line of infantry around the agency buildings, dug some rifle pits, put up barbed wire in places, and set a guard about the big government boarding school in which over one hundred boys and girls were being kept, partly to protect them from the ghost dancers, partly as hostages for the good conduct of their fathers, who were mostly chiefs and headmen. The Sioux passed through the lines freely in daytime, but no one was permitted to enter the agency enclosure after 4:30 in the afternoon.

The Brûlés of Rosebud had never forgotten that terrible day in 1855 when Brigadier General W. S. Harney's troops had caught them in camp on the Bluewater in western Nebraska and had administered to them such a beating as no Sioux had ever before experienced. Even in 1890 any report that soldiers were coming was sufficient to set the Brûlé women to wailing and the children to screaming. Now, in mid-November, 1890, many of the Brûlés did not have very good consciences. They had been flouting the agent and his police. The Wazhazhas and the former Sitting Bull hostiles settled in the northwestern part of the reservation were openly indulging in ghost dancing and were also killing large numbers of stock cattle to make feasts in the ghost dance camps. Two Strike and his Brûlé Band Number Two were hesitating as to whether they should join the ghost camp or not. At this critical moment a mixed-blood employee at the agency sent the Indians a secret warning that troops were coming. The alarm spread like fire. Two Strike's band was swept into the ghost dance camp and was followed by Long Mandan and his Two Kettle band who fled from their camp near the location where the town of White River was later built. Some progressive families fled in the opposite direction, to the agency, trusting their white friends to protect them from the troops.

Short Bull, the ghost dance prophet, now had in his camp not only the former Sitting Bull hostiles but most of the Wazhazha Band, most of Two Strike's people, the Two Kettles, and the Bull Dog Sioux (so called from their leader, Chief Bull Dog). Many of these Sioux were not ghost dancers. They had been brought into Short Bull's camp by panic fear of the troops, and when their fear subsided they found themselves "coralled" by the ghost dance fanatics, who threatened to kill them if they attempted to leave the camp. Old Two Strike seems to have doubted the wisdom of remaining in the ghost dance camp; but Crow Dog was at his side, advising him to remain. Two Strike was a fighting man with little brain. He listened to the wily Crow Dog. Besides, part of his people had gone over to the ghost dance faith, and he now could not leave without splitting his band to pieces.[10]

[10] Two Strike was born in 1821 and died in 1914. He was a stanch supporter of Spotted Tail up to 1880, when he listened to Crow Dog and other plotters and was induced to side with them, at least for the moment. When Spotted Tail was murdered by Crow Dog, Two Strike was left without a leader and drifted this way and that. He remained a pagan and nonprogressive; but his poor mental equipment caused him to make foolish decisions, which cost him standing in the tribe and lost him many old followers.

These events had put a majority of the Rosebud Sioux in the power of Short Bull. He and his armed fanatics assumed control of the mass of frightened Sioux, and an order was issued for everyone to prepare to move westward to Pass Creek. Everything played into the hands of the ghost dance leaders. They had planned the move to Pass Creek in October, and now it fitted in with the desires of all the Brûlés, whose one wish was to get as far away from the troops at the agency as possible. Short Bull had ordered the Indians to abandon all earthly goods and prepare for the coming of their ghost relatives in early December; but the Sioux were not mad enough to try to go naked and live on air in a Dakota winter. They did abandon their log cabins and much of their belongings, but they took heavy clothing, they loaded their wagons with their families and some necessary articles, and they drove their ponies and stock cattle along when they started for Pass Creek. Thus the last great migration of the Brûlé Sioux was begun in mid-November, 1890.

II

Touch and Go

GENERAL BROOKE found Red Cloud, Red Shirt, American Horse, and some other chiefs awaiting him when he reached Pine Ridge agency. Red Cloud was now almost blind and could no longer take an active part in affairs. He and his friend Red Shirt were on the fence, not taking sides. Jack Red Cloud had sided with the ghost dancers; but he had not the determination and character of his father, and the minute the troops appeared he began to pose as a friendly. The other chiefs in camps near the agency—No Water, Four Bears, Yellow Bear and He Dog—were openly siding with the ghost dancers. None of the ghost dance camps were within reach of the troops.

General Miles had no purpose of taking offensive action. He had sent troops to protect the agencies and he was throwing a screen of troops around the Sioux reservation, just in case some of the ghost dance fanatics might develop a desire to leave the reservation and raid the white settlements. This was a wise and humane policy, but it was expensive, as large forces had to be kept in the field, probably for months. The crisis could only end with the return of the ghost dance fanatics to a rational state of mind, and no man could say when that would be.

General Brooke found Agent Royer's wildcats at Pine Ridge rather tame. They grew tamer when he assured the chiefs that the troops would not harm the Sioux in any manner and that this applied even to the ghost dancers. The agency chiefs smiled; but from the distant ghost dance camps came only yells of defiance. When the prophets had started the ghost dance in early summer they had done so under the guise of peaceful assembly, with a rule that all Sioux should come to the dances without arms; but now in mid-November they danced with rifles strapped on their backs and their naked bodies crisscrossed by belts of cartridges. They kept scouts out in every direction, watching from the hills and ready to spread the alarm instantly if troops were seen on the march.

The Pine Ridge ghost dancers were all in the east, at Big Road's camp of former hostiles on Wounded Knee and in Little Wound's progressive settlement on Yellow Medicine Root. On November 20 news was received that Short Bull with the Brûlés from Rosebud was marching into the eastern borders of Pine Ridge. Thus the two main masses of ghost dancers were united, and to keep the Sioux at the agency quiet, General Brooke on November 27 ordered Agent Royer to ignore the beef cuts ordered from Washington and to restore the Sioux rations to the full treaty quantities. In Washington, Commissioner of Indian Affairs Morgan had claimed for months past that there was no way to increase the amount of food for the Sioux except by waiting for an act of Congress. General Brooke made the necessary alteration in five minutes. The Sioux at the agency, getting full rations again for the first time in years, began to think that white soldiers were not such bad people. General Brooke pleased the Sioux very much by setting aside Commissioner Morgan's ladylike rules for butchering and letting the Indians go back to the old method of pursuing the cattle on horseback and shooting them with rifles.

A whole generation had grown up since the Sioux had come on the reservation, and during the whole of this period the leaders of the Eastern Indian welfare groups had been preaching that chiefs must be destroyed and every Sioux must learn to think and act for himself. Now in 1890 all this white-man nonsense had been blown away on the wind. It was to their old chiefs—Red Cloud, Little Wound, Man-Afraid-of-His-Horse, Two Strike, and the others—that the Sioux turned for guidance in this time of trial; and it was due to the advice of some of these chiefs that all of the Sioux did not flee into the badlands when the troops appeared. As for the younger generation that the Eastern Friends of the Indians were always terming unctuously the "hope of the new day," it would be difficult to name a half-dozen outstanding young Sioux who were friendlies in 1890; but a surprising number of young men of promise had turned their backs on civilization and were among the most dangerous ghost dancers. With them were several returned students from Carlisle—the ewe lambs of the Eastern humanitarians, who believed that education would quickly eradicate all traces of *Indianism*. Was not the ghost dance pure *Indianism?*

The coming of the troops assured quiet at the agencies, but there was no method for dealing with the ghost dancers without the risk of bringing on a general Indian war. General Miles and the other military leaders knew this. Conditions were very different from what they had been when the Sioux had first come to the reservation. At that time the white hunters were finishing the killing of the buffalo herds; the Indians had lost their means of support by hunting and had to accept the offer of reservations. Now in 1890 the old buffalo range had been restocked with cattle, and if the Sioux and other Plains tribes broke away from their reservations they could live for a long time by raiding the cattle herds and white settlements. Joined by other tribes, the Sioux might spread war through the plains from Canada to Texas before our small military forces could control the situation.

But in late November, 1890, none of the ghost dancers seemed to have any plan beyond holding out on the reservation until their ghost relatives came from the spirit world to join them. They had the promise from their prophets that the whites would be destroyed in a great dust storm. There was no need to fight them. But there was the matter of food. The ghost dancers had cut themselves off from the free rations at the agencies. They were living by stealing stock cattle on the reservation and by raiding the slender supplies of corn and vegetables some of the farming Indians and squawmen had stored up. When these sources of food were exhausted, the time of real danger would come; for the Sioux would not sit in their camps and starve with the opportunity at hand to obtain all the food they required by leaving the reservation and starting raiding. General Miles knew this; but he knew another thing. That was that some of the ghost dance leaders, old chiefs like Little Wound and Two Strike, were not completely convinced by the teachings of the ghost dance prophets. They had their doubts about the return to earth of the Indian ghosts and the destruction of the whites in a dust storm, and they were soon considering the advisability of gathering up their followers and deserting the ghost dancers. General Miles was already trying to get in touch with these chiefs in the ghost dance camps, hoping to draw them and their bands to the agencies.

Little Wound was still in his own settlement on Yellow Medicine Root Creek when Short Bull and all the Rosebud ghost dancers arrived on Pass Creek. By this time all the Sioux families that doubted the ghost dance beliefs had left their log cabins and gone to join the friendly camp at Pine Ridge agency, abandoning their property and leaving their stock cattle to shift for themselves. If there had been an agent he could trust and if the troops had not come, Little Wound might have gone to the agency. But he detested Agent Royer, and he had a clear memory of the day twenty years back when the cavalry had swept down on his camp and driven him and his people in wild flight north across the Platte, and he had a profound mistrust of the army. The white soldiers never came except with the purpose of starting a fight. This was the view held by most of the Sioux, and the coming of Custer's old regiment to Pine Ridge had at once started a rumor that the Seventh had come to seek vengeance on the Sioux for the killing of Custer and his men in 1876.

For a short time the increase in beef and other rations made the Sioux at Pine Ridge agency happy; but soon they began to be alarmed again. They could not understand why the troops sat down at the agency to wait. Was it a plot, to wait until the Sioux were off their guard and then attack them? Or were the whites afraid of the ghost dancers? Short Bull from Pass Creek sent scouts westward, and these parties rounded up the stock cattle which the friendly Sioux had abandoned when they fled to the agency. The ghost dancers also stripped the Sioux cabins and carried off the plunder to their camp. Many of the Sioux of the Little Wound and American Horse settlements were now Christians. They had been taught by their missionaries to make the interiors of their log cabins neat by lining the walls with the canvas which the government issued to the Sioux, primarily for making tipi covers. These Christian families in the friendly camp were very much shocked to hear that the ghost dancers were cutting the canvas from the walls of their cabins and using it to make ghost shirts, painted with pagan symbols which were supposed to protect the wearer from bullets. The worst of it was that, with the government apparently helpless in the face of the ghost dancers' activities, many of the Christian Sioux began to wonder if the ghost dancer prophets were not telling the truth. As Sioux Indians these Christians were perturbed about what their fate would be if the white race was really destroyed.

The progressives in the friendly camp were also in deep trouble. They had followed the advice of the agents and tried to farm and get on; but now they had lost everything. American Horse was an example. He had been a wealthy chief. He had gone traveling with Wild West shows and had acquired many fine things. He had a nice cabin, herds of ponies and cattle, and now the ghost dancers had cleaned him out and he had nothing left but a canvas tipi in the friendly camp. He could not even be safe there; for the ghost dancer sympathizers were threatening to kill him, and he had to spend most of his time with the troops at the agency.

Even the military at Pine Ridge now became much perturbed. Was this policy of keeping the troops inactive and trusting that time would cool the ardor of the ghost dancers producing anything beyond a growing threat of final disaster? The ghost dancers seemed to be recruiting and concentrating their strength every day that the troops sat idle. General Miles answered this growing threat by sending more troops to the agencies. At Pine Ridge they marched in about December 1, and at the same time former agent McGillycuddy arrived, serving as official observer for the governor of South Dakota. His appearance at the agency was the signal for his friends, who at once started pelting the officials with demands that he be put in control again of the Pine Ridge Sioux. No reply came from Washington, and General Brooke exhibited no inclination to make use of McGillycuddy in any official capacity.

Just how serious the situation appeared at this time was disclosed when General Miles on December 1 stated that this was the most dangerous Indian crisis since the days of Tecumseh, the Shawnee prophet. Some Sioux had gone northward even into Canada and others south to Indian Territory, spreading the messiah craze among the tribes they visited. This seems to have been the casual action of individual Sioux who simply wandered off to visit friends among other tribes; but at the time it looked like a plot of the Sioux leaders who were proposing to bring on a general Indian war. General Miles stated that our army had only two thousand cavalry and that the infantry was useless in the pursuit of mounted Indians, if the tribes should break away from the reservations and start raiding. But there was comfort in the facts that were slowly developing and that showed the Sioux had no plans for taking the offensive, no organization, and no leadership of any quality. The one leader who was dangerous was Sitting Bull, and if any of the Sioux had a plan for offensive warfare, he was the man.

On December 4, Hollow-Horn-Bear and one hundred leading Brûlés of Rosebud put their marks to a petition in which they asked the President for food for their starving families.[1] These were the friendly Brûlés who had gone to the agency when the ghost dancers had fled from Rosebud westward into the Pine Ridge lands. At this same date the Brûlé ghost dancers, having consumed all the food to be found in the Pass Creek and Yellow Medicine Root districts, moved their camp westward. They raided the settlement of squawman and mixed-blood families at the mouth of Porcupine Tail Creek, robbing these people of their cattle, horses, and other property. Pressing onward, they captured the Pine Ridge beef herd and the herd camp near the mouth of Big White Clay Creek. Here they crossed White River and encamped in a very strong position in the edge of the Big Badlands, where it was reported by scouts they had five hundred women at work digging rifle pits.

The ghost dancers had parties of scouts out in every direction, extending southward to the vicinity of the Roman Catholic mission, only four miles north of Pine Ridge Agency. They were so

[1] Boyd, *Recent Indian Wars*, 233.

watchful and in such a vicious temper that it was impossible for the best government scouts and even for friendly Sioux from the camps close to the agency to penetrate the screen of scouting parties and learn what was going on in the ghost dance camp. It was Father Jutz, aided by Red Cloud and some of the other chiefs at the agency, who now succeeded in coaxing some of the chiefs among the ghost dancers to come to the agency for a conference with General Brooke. With all the friendly camps close to the agency seething with excitement, the ghost dancers marched in on December 7. They came in style, a mounted warrior ahead carrying a white flag of truce; behind him the Indian soldiers, painted and befeathered and riding on painted war ponies whose tails were tied up with ribbons and ornamented with eagle feathers. This was one of the last real war parades of the Sioux and the throngs of watching Indians went wild with excitement. These Indian soldiers who led the march—Turning Bear, Big Turkey, High Pine, Big Bad Horse, and Bull Dog—all had Winchesters held in the crooks of their arms and were ready to fight at an instant's warning. Behind them rode a group of chiefs on their ponies, wearing fine Sioux war costumes and (a new touch) the fantastic ghost shirts and ghost leggings. At the tail of the procession rode Father Jutz and Chief Two Strike of the Brûlés in a plebeian and dilapidated buggy with a mounted guard of four young Brûlé warriors to protect them. Riding boldly in among the troops, these Sioux tied their horses to trees near General Brooke's headquarters and entered the building for the council.

This council produced little in the way of immediate results; but it did exhibit the fact that Two Strike and some of the other leaders were anxious to come to the agency with their bands and that a large part of the ghost dancers were hesitating on the verge of such a decision. The Sioux were no fools, and the majority of the men in the ghost dance camp were beginning to suspect that the prophets who had led them into this strange venture were either fools or rogues. The ghost dance fanatics were giving themselves away. To convince waverers that the troops really could not hurt them, they had put an allegedly bullet-proof ghost shirt of cloth on the fanatic named Porcupine and had permitted several warriors to fire at him with Winchesters. Badly wounded in the hip, Porcupine had lost faith, and so had most of the spectators. To induce some half-hearted followers to desert their warm log cabins and live in tents in the ghost camp, the prophets had declared that a miracle was to be performed; there would be no winter in the Sioux land, it would be warm all the time from October to April. Now winter had come down on them at its usual time and they were shivering in tipis made of the cheap and flimsy canvas the government had issued to them. They had eaten most of the cattle within reach and were beginning to be very hungry. Therefore they were hoping that Two Strike and the other old chiefs would come to terms with General Brooke. But this was being violently opposed by the fanatical ghost dancers, who threatened to shoot any Sioux who attempted to leave the ghost camp. The fanatics jeered at the idea of trusting the word of a white soldier-chief. Thus, afraid of the ghost dancers and equally afraid of the troops, the majority of Indians in the ghost camp hardly knew what course to take. This was clearly exhibited in the council with General Brooke. Turning Bear spoke first, as he was the leading ghost dance soldier. His tone was friendly, but he was using putting-off talk. Yes, he said, they would like to come and camp near their friend General Brooke; but all these Sioux (the friendly camps) near the agency made the land too crowded. There was no grass or water for the ghost dancers' great herds of ponies. Besides, they had many old people in their camp who could not ride, and they had no wagons for these poor old folk to be brought along in. Two Strike, who was the biggest man in this party, was clearly anxious to bring his band to the agency; but he was being held back by the opposition of the ghost dance fanatics. General Brooke was quite conciliatory in his talk with the Indians, and when the council ended he presented them with a number of boxes of army hardtack and other rations. After two hours of talking the party left the agency, heading north toward the ghost camp.[2] They went home to discuss among themselves this matter of moving to the agency, and their councils were not

[2] *Ibid*, 238. See also Colby, in *Nebraska Historical Society Transactions*, III, 150ff.

quite as good-natured and polite as had been their conversations with General Brooke. Kicking Bear, the Cheyenne River prophet, was now in the ghost dance camp aiding the other prophet, Short Bull, in keeping the Sioux in paroxysms of religious excitement and urging them to have nothing to do with the whites at the agency; but General Brooke now had his own representatives in the camp. Louis Shangraux, a brave Sioux mixed blood, accompanied by Chief No Neck and thirty-two young men from the friendly camps, had boldly pushed into the badlands fastnesses and entered the ghost dance camp, to urge the chiefs to bring their people to the agency.

When this party entered the camp, they found a wild ghost dance in progress. It was kept up without a halt for thirty hours, and then was recessed only because some of the chiefs wished to talk with Shangraux and No Neck. In this council Short Bull the prophet, Two Strike, and Crow Dog acted for the ghost dancers; but it was Short Bull who summed up for his side in a violent harangue in which he stated that his followers would not give up the ghost dance or move to the agency, that they did not trust the whites, especially as they feared that once encamped at the agency the troops would disarm them and take their ponies. They would then be put in jail for stealing stock cattle and plundering the log cabins of the progressive Sioux. He concluded by telling Shangraux and No Neck to go back to their white friends at the agency with this message. But Shangraux and No Neck with their young warriors stuck. The ghost dance was now resumed and was kept going for two days and nights; then, on Saturday, Two Strike suddenly announced that he was going with his band to the agency. A council was called and Crow Dog backed Two Strike; but Short Bull sprang up and shouted furiously that these men from the agency were lying, that any men from the ghost dance camp that ventured to the agency would be at once put in jail. "Louis is at the bottom of this!" he shouted. "I know that he is a traitor! Kill him! Kill him!" A group of young fanatics clubbed their guns and rushed at the Sioux from the agency, who were forming a screen around Shangraux, No Neck, Two Strike, and

Crow Dog. The Indians were all yelling and cocking Winchesters and a general fight seemed about to be precipitated, when that consummate actor Crow Dog drew all eyes on himself and caused a sudden hush in the fierce shouting by sitting slowly down and covering his head with his blanket. He cried out that he could not bear to see the Sioux shed Sioux blood. He waited for that to have its effect, then removing the blanket from his face, he boldly fronted the ring of armed ghost dancers. "I am going back to White Clay,"[3] he said. "You can kill me if you want to now, and prevent my starting. The agent's words are true, and it is better to return than to stay here. I am not afraid to die." Thus the man who had shed Sioux blood without compunction in the killing of Spotted Tail and who had talked Chief Two Strike into bringing all his people to the ghost dance camp prevented a general fight at this critical moment and turned the scales against Short Bull and his ghost dance fanatics. On this day the Crow Dog who was a malicious plotter and troublemaker was not there; the Crow Dog who might have been a great leader among his people was.

Louis Shangraux, No Neck, and their scouts now left the ghost dance camp, taking with them Chief Two Strike and Crow Dog with about one hundred lodges. Up to the last moment the ghost dance fanatics continued to yell fiercely and brandish their weapons, but not a shot was fired; and when the marching Sioux reached a low ridge two miles from the camp and turned to look back, they were amazed to behold the ghost dancers frantically taking down their lodges in the midst of wild confusion. Presently over one hundred lodges of ghost dancers were advancing across the plain toward the ridge where Two Strike's people stood waiting. It now became known that the sight of Two Strike's camp starting for the agency had proved too much for the jumpy nerves of the Sioux who were being left behind, and—brushing aside the fierce protests of the Short Bull fanatics—the people had decided to go with Two Strike. Even Short Bull came along; but after the united camp had advanced another four miles toward Pine Ridge, he became suddenly alarmed and turned back into

[3] Pine Ridge Agency.

the badlands with a few followers.[4] Quickly informed of these events by Sioux scouts, General Brooke asked the chiefs at the agency to send two hundred of their men to join Two Strike and to urge him to turn back and attempt to coax Short Bull and the other fleeing ghost dancers to rejoin the main camp. With these two hundred Oglala warriors from the agency went Merrivale, an Indian trader of French blood. That night a number of military officers and other white men were dining at Asay's trading store at Pine Ridge, both Agent Royer and formed agent McGillycuddy being present. In the middle of the meal Royer was called to the door to receive a message, and a moment later he rushed back shouting, "It's come! War's come! The hostiles wouldn't listen to Merrivale! They shot over his head, killed a lot of cattle, and lit out for the badlands!"[5]

For weeks the nation had been watching in wonder and amazement the drama of this ghost dance movement as it was unfolded from day to day. What on earth were these Indians up to? Here we stood, convinced that our country was the most progressive and civilized in the world; here we stood on the threshold of the great new twentieth century that everyone was talking about, waiting eagerly for the inventors to produce the first horseless carriage, scanning the skies at night whenever it was rumored that at last a real airship had been built and was making secret trial flights under cover of darkness, and here were the Sioux—men, women, and children—holding hands and dancing naked in the snow in eager preparation for the destruction of the whites and the return to earth of the ghost Indians and ghost buffalo. Ghosts, indeed! And ghost warriors (very much alive) defying the entire United States Army in perfect faith that their ghost shirts would protect them from wounds!

[4] Boyd, *Recent Indian Wars*, 209. The date was apparently December 12. On that day Brooke reported Two Strike was moving to the agency and that Short Bull had fled back into the badlands.

[5] *McGillycuddy Agent*, 263. *Nebraska Historical Society Transactions*, III, 150. Boyd, *Recent Indian Wars*, 246. Merrivale was the brave French trader who went to the hostile camp with Spotted Tail's party in late winter 1876–77. He was a friend of Crazy Horse and other hostile chiefs. He stayed in the camp after Spotted Tail left and helped bring the hostiles to the agencies to surrender in April, 1877.

These were dreadful days for the men who had controlled our Indian policy in the assured faith that, since they called themselves friends of the Indians and pretended to understand them, they had a kind of divine authority to meddle. In the Senate, Senator Henry L. Dawes of Massachusetts found himself in a most uncomfortable position. For almost ten years he had let himself be used as the Senate leader for the benevolent societies who were attempting to control Indian policy, and he knew better than any other man that this condition among the Sioux was no mere accident. It was not even wholly the unexpected result of Congressional economy and the systematic paring down of Sioux appropriation bills with the object of saving money. It was largely due to the deliberate planning of good Christians in the East, who had taken it upon themselves to direct the lives of the Sioux, even to the extent of exerting pressure on these Indians through hunger. Had not their leaders quoted as applicable to the Indians what they termed St. Paul's dictum; "He who will not work shall not eat," and had not the brethren approved that doctrine and carefully laid plans to cut rations and thus apply the necessary pressure to force the Indians to progress? Who could have foreseen the shocking reaction of the Sioux to this well-meant Christian attempt to starve them and their children, for their own good? Instead of meekly submitting, these Indians had put on ghost shirts and killed government cattle on the reservation, feasted and prepared themselves to die a good old-fashioned Sioux death in battle! In the Senate, Dawes grimly prepared to defend the Indian policy of these Christian brethren, for which he had made himself largely responsible.

The debate was precipitated by the introduction of a bill on December 3, 1890, providing for the issuing of 100,000 stands of arms to the citizens of Dakota, who were pictured as in hourly danger of murder although not one Sioux had left the reservation or exhibited any desire to attack the whites. Senator Voorhees mildly suggested that, since the Sioux trouble was clearly the result of hunger among these Indians, it might be better to send 100,000 rations to the Sioux rather than the same number of rifles to excited border whites. Another Senator brought up the current gossip, that the Crook land commission of 1889 had made many handsome promises to the Sioux—among others a promise of more beef—which the administration and Congress had failed to honor, and that General Crook's death in the preceding March had been mainly due to grief over this breach of good faith. Senator Dawes angrily denied that the administration or Congress had failed in any way to carry out General Crook's promises. He stated that he realized the danger in the present Sioux situation, but that no relief was to be expected until these Indians were removed from the leadership of chiefs like Sitting Bull and Red Cloud, who were the curse of their race. At the moment he spoke these angry words Red Cloud was the chief mainly responsible for holding nearly one thousand lodges of Sioux in the friendly camps at Pine Ridge, and this he had accomplished through the respect and honor his people paid to him, for he was now too blind to be really active. As for Sitting Bull, his following in 1890 was very small and purely local, and it was absurd to term him a leader of all the Sioux.

Senator Voorhees asked if Dawes would state whether General Miles was correct in attributing the Sioux troubles to hunger. Dawes replied tartly that this day he had heard for the first time of hunger among the Sioux. These Indians, he angrily stated, had been ghost dancing and "going off on the war path" for weeks past, and now this talk of starvation had come up. He had no doubt that if these Indians had stayed at home on their farms they would have had sufficient food. Voorhees asked if the Senator did not know that this condition of hunger went back at least two years. Dawes burst out in an angry tirade. The Sioux were hungry, he declared because they had left the agencies, had abandoned their tipis and belongings, to be plundered by the class of their people who never did a day's work—six thousand loafers—the wild, wicked *Brûlés* and *Kiyuksas!*

And this was the man who was the great authority in the Senate on Sioux affairs—the man who had led the fight in Washington for the right of the brethren to control the destinies of the Sioux! He did not even understand the simple fact that, until the ghost dance drove them mad, the Brûlés had been rated the most friendly of the Sioux and that the Kiyuksas (Chief Little Wound's band at Pine Ridge) were good Episcopalians and, as far as the Dakota climate would permit, they were leading in the attempt to support themselves through farming. To picture these particular Sioux groups as wicked "won't works" was a sad exhibition of ignorance on the part of the speaker, and Dawes did not increase his reputation for levelheadedness by his words during this debate. He must have felt very queer when General Miles wrote (December 19) that the roots of the Sioux trouble lay in "the forcing process" by which the government had attempted to compel the Sioux to win self-support from farming at a time when the Indians could not grow crops because of the great droughts.[6] And who had thought up that forcing process and taken it to Washington as a good and worthy policy that Christians who were interesting themselves in Sioux welfare could heartily recommend? The brethren of the benevolent societies had done that, and Senator Dawes, their principal spokesman in Congress, had worked to have that policy put into force. Miles gave as another cause of the troubles the fact that the Sioux had been coerced into signing away their lands in 1889; and who had planned that campaign? Again, it was the brethren, with Senator Dawes aiding them to the full of his ability.

As we have seen, when Two Strike broke away from the ghost dance camp in mid-December and took his band of Brûlés to Pine Ridge Agency, the ghost dancers—seething with wrath and frantic with alarm—fled north across White River, taking with them the government beef herd they had stolen. They took refuge on a high dry terrace or plateau in the Big Badlands north of White River, a place called the Wall Camp; and from this stronghold (full of good beef and in a bellicose mood) their leaders loudly defied the United States Army to come and fight them. But at Pine Ridge there was a feeling of confidence that these wild fellows did not really mean it and that they could be coaxed into coming to the agency. Miss Emma G. Sickles, the New England spinster who was head of the Pine Ridge boarding school, took the opposite view and busied herself in spreading alarmist reports that a dark plot was being hatched by Red Cloud and "the treach-

[6] Miles letter, in Mooney, *The Ghost-Dance Religion*, 835.

erous ones." While she talked, Red Cloud and the other chiefs (December 20) sent five hundred picked warriors from the friendly camps, led by Jack Red Cloud and other younger chiefs, into the badlands to make another effort to coax the ghost dancers in. They came back to the agency with a considerable body of deserters, but the most fanatical of the ghost dancers refused to leave their stronghold.

In mid-December every prominent Oglala and Brûlé was in the friendly camp at Pine Ridge, aiding the government's effort toward a peaceful settlement. Red Cloud and American Horse, who had been far from friendly, were now allies. Jack Red Cloud, who had made a false start in summer by siding with the ghost dancers, had been talked around by his father, and the newspapers were referring to him as "Jack Red Cloud, the highly-respected son of the great chief." Two Strike of the Brûlés and Little Wound and Big Road of the Oglalas (really big chiefs) had deserted the ghost dancers, and even Kicking Bear, the ghost dance prophet, was now in the friendly camp.

These were the first Indian troubles during which the telegraph and the railroad brought modern facilities up to the front lines, and cranks as well as troops had taken the field. One December night Red Cloud's own camp, just across the creek west of the agency, was thrown into intense excitement by the appearance of a rural Iowan named Hopkins who had been mixing Bible study with botany and some peculiar brand of mysticism, producing a new religion which he termed the faith of the Star Pansy Banner. Dressing himself up like a wooden cigar store Indian he had come out on the railroad to a station a few miles south of Pine Ridge, and now he popped up suddenly like a magician in the middle of Red Cloud's camp, announcing himself as the new messiah. He could not have picked a worse spot for making his debut as a prophet. This mass of Sioux, taken from isolated farm communities and crowded together in canvas tipi camps (one thousand lodges), was in a highly nervous condition. For weeks and almost hourly wild reports had spread through the camps, throwing the Indians into violent excitement of either alarm or rage; now when the star pansy chief made his mysterious appearance and announced himself as a messiah, some of the Sioux accepted him as the savior who had been long reported on the way to help them, but the majority in Red Cloud's camp declared that he was an imposter, and before he had time to explain the purpose of his mission the shocked star pansy was being violently yanked this way and that by contending factions of enraged Sioux, who were all yelling and handling cocked Winchesters with a hair-raising disregard for possible consequences. Rescued by a little party of Indian soldiers and mixed bloods, the frightened prophet was led before Red Cloud. The old chief listened to explanations and found himself in another dilemma, of which he had had to face far too many in recent weeks. Now he was suddenly called upon to act the part of Pontius Pilate and to pass judgment on a messiah. As a Catholic convert he knew that the whites had once had a true messiah, and perhaps he was now face to face with one who had been sent to the Sioux; but Red Cloud had plenty of hard sense, and after questioning the Iowan through an interpreter, he saw clearly that the man was only a fool. "You go home," he said severely. "You are no son of God." "Just a —— —— son of a ——!" shouted an irate squawman, who had nearly acquired a skinful of Winchester bullets while helping to pluck the Iowan out of the center of a mass of raging Sioux. Spirited away from the agency at night, Messiah Hopkins was taken under military guard to Rushville on the railroad in Nebraska, where he was put on a train headed toward Iowa. Recovering from his fright, this man bombarded the officials in Washington with excited letters, in which he maintained that he was the true Indian messiah and that the only hope for improving conditions among the Sioux rested in official approval of his mission to go among these Indians and preach the true faith of the Star Pansy Banner.[7]

As none of the Sioux leaders exhibited the least inclination to start a war, all might still have ended in the ghost dancers coming

[7] A few old men at Pine Ridge still remember the Star Pansy prophet. General Colby has an account of this man in his history of the campaign in *Nebraska Historical Society Transactions*, III, 154. Hopkins signed his letters *Albert C. Hopkins, Pres. Pro. Tem. Pansy Society of America.* His original idea seems to have been that the millenium would arrive when the wild star pansies bloomed in the Iowa prairies in April, 1891.

quietly to the agencies if it had not been for the stubborn conduct of Sitting Bull. This Sioux had hated the whites and all their ways since his youth, and he had made trouble for the government from 1864 on. He had led in the Sioux war of 1876; yet when he returned from Canada in 1881 and surrendered, he was treated with great generosity, he and his leading men and their families being kept in a pleasant camp at Fort Randall for a few months, after which they were sent to the Standing Rock reservation. But Sitting Bull was incurable, and from the moment he was released from military supervision he attempted to put himself at the head of a group that stood for Sioux paganism and, one is tempted to add, for something like open hostility toward the government that was feeding and caring for them. The Sioux all knew about this, and what they thought about it we may judge from the fact that when at this period they wished to name a chief as head of their nation, they sent delegations to Pine Ridge to ask Red Cloud to accept that position, passing over Sitting Bull's claim and leaving him the rank of chief of a small group of former hostiles at Standing Rock.

In June, 1890, the officials at the Indian Office were considering arresting and removing Sitting Bull from the reservation; but they seemed to desire to place the responsibility for the arrest and its possible consequences on the shoulders of Agent McLaughlin. The agent, warned by his canny Scotch blood that the instructions sent to him were more in the nature of recommendations than downright orders, refrained from any action further than keeping a watch on Sitting Bull and reporting frequently to the Indian Office. The officials now suggested to Brigadier General Thomas H. Ruger, commanding the troops on the Missouri, that he should send out a force to arrest Sitting Bull; but Ruger was no more willing to burn his fingers than McLaughlin was. Indeed, he promptly passed the hot penny back to that agent with the remark that it would add greatly to the prestige of the Standing Rock Indian police if they, rather than the troops, made the arrest. McLaughlin continued to write reports but took no further action.

When the military were put in control on the Sioux reservation in November, General Ruger acted first against Chief Hump of

Courtesy Bureau of American Ethnology

Sitting Bull

Courtesy National Archives

Big Foot's band at the Ghost Dance at Cheyenne River, August, 1890.

Cheyenne River. This Mineconjou chief was regarded as potentially even more dangerous than Sitting Bull. In 1876 Hump had led the Sioux into battle against Crook and Custer and since coming to the reservation he had clung to old Sioux customs, often openly defying the agent. General Ruger acted wisely, sending Captain E. P. Ewers, who knew Hump intimately and was trusted and liked by that chief, to deal with him. Captain Ewers came all the way from Texas to perform this duty. On reaching Fort Bennett at Cheyenne River Agency, he selected Lieutenant H. E. Hale as a companion, and the two officers rode sixty miles up Cheyenne River through country considered too dangerous for any white man to venture into, to Hump's camp at the mouth of Cherry River. Hump was twenty miles away in another Sioux camp when word was brought to him that the officers were seeking him. He might easily have slipped away; but he was no fanatic like the ghost dance prophets, and he decided to ride at once back to his own camp. After a talk with Captain Ewers he agreed to bring his whole band (some four hundred people) to the agency, which he promptly did, and he and many of his men enlisted as scouts to act against the ghost dancers.[8] It was hoped that through Hump's influence Big Foot, whose camp was farthest from Cheyenne River Agency, ninety miles west at the forks of Cheyenne River, could be induced to come in. That was the last of the ghost dance camps at this agency. Although the ghost dancers at Cheyenne River had defied the agent and his police for months past, they had not committed one act of hostility against the whites.

Dealing with Sitting Bull was a different proposition. The man was a fanatic, or if you choose a mystic, whose actions could not be foreseen. From June, 1890, onward, every possible plan was considered, but the officials always came back to their first idea of having Agent McLaughlin's Indian police deal with Sitting Bull. McLaughlin was ready to act but only on definite orders, and such orders the Indian Office officials seemed unwilling to give. When the military took control, the situation was altered, and now an amazing order came from army headquarters, that

[8] *Report of the Secretary of War* (1891), I, 147.

W. F. Cody (Buffalo Bill) should go to Standing Rock Agency and arrest Sitting Bull. Why Cody was chosen for this extremely delicate piece of work is difficult to determine. He was a fine showman; he had been a scout in the Sioux war of 1876, and just when his services were needed with the troops in the field, he had gone East to fill a stage engagement. With a genius for showmanship, he had induced Sitting Bull after his surrender in 1881 to join the Cody Wild West show and had exhibited him about the country. In 1890 he was posing as Sitting Bull's close friend. Perhaps that was why he was selected to deal with the chief.

Agent McLaughlin was quite evidently taken aback and very angry over this proposal to bring Cody into the delicate situation at his agency. He had in his employ scouts who had seen twenty times the service among the Sioux that Cody had and who knew Sitting Bull as one neighbor knows another. They also realized the kind of desperate fight that might ensue on any effort to arrest the chief, and they were prepared to face that possibility boldly and skillfully. Cody's Indian fighting had been mainly a matter of riding rapidly up and down, exchanging shots with Indians who were also riding rapidly up and down at a range of some hundreds of yards. What would he do on entering Sitting Bull's log cabin, if he ever got that far, and finding himself faced by ghost dance fanatics, the muzzles of their Winchesters actually touching his body? That would not be a fight but more like suicide—an affair in which he might get himself killed; but Sitting Bull would probably escape and summon the Sioux to battle.

Cody made his preparations with a showman's flourish and came to Standing Rock with some cowboy associates; but Agent McLaughlin made stiff protests to Washington and the President himself intervened, ordering that Cody should leave at once. The responsibility for dealing with Sitting Bull was now back where it had been before, on the shoulders of General Ruger; but instead of acting, Ruger began to plan a military campaign along the lines of that of 1876. Exactly as Crook and Terry had planned to strike the hostiles in winter, Ruger now talked of waiting until cold and snow pinned the Sioux down in their camps, and then he would send troops to pick up Sitting Bull. But by December 5 it began to seem very doubtful that the Sioux would sit where they were, patiently waiting for winter to come to General Ruger's aid. McLaughlin had his Indian police and some spies watching Sitting Bull and even visiting in his log cabin camp on Grand River, and these men reported that the chief was preparing to move, perhaps to flee to the Big Badlands and there put himself at the head of the ghost dance camp. That would constitute a pretty kettle of fish for the military to deal with.

General Ruger appears to have been the weak link in the chain that General Miles had drawn around the Sioux reservation at this time to prevent any of the Indians from breaking away. Ruger had a fine Civil War record, but he did not seem to know how to deal with Indians and he had little sense of the value of time, which was now of first importance. The moment had struck; but he was still indulging in the luxury of changing his mind and at the moment his favorite scheme was to let the Indian police unaided deal with Sitting Bull, as the prestige of the police had been badly damaged during the summer and fall and needed bolstering up. He seemed unable to realize that he had not been put in charge of an operation intended to improve the prestige of the Indian police, but of one whose sole object was to secure the person of Sitting Bull and to accomplish that, if possible, without bloodshed.

After all the final decision was forced on the officials in Washington, who had tried so long to fub off the responsibility on Agent McLaughlin or General Ruger. A meeting of Indian Office and War Department officials was held and definite orders were sent to Agent McLaughlin to act in unison with Colonel W. F. Drum, commanding the troops at Standing Rock Agency, and make the arrest at once. McLaughlin had his plan and men ready. On December 19–20 his Sioux would come to the agency for their rations—all except Sitting Bull's followers, who had not ventured near the agency for months because they feared arrest. With Sitting Bull thus isolated far away on Grand River, McLaughlin would strike with his Indian police supported by Drum's cavalry and have the business wound up before any of the Indians assembled at the agency could come to Sitting Bull's aid. It was an excellent plan; but the wily chief seems to have thought along the same lines, concluding that while all the Indians and whites were busy at the agency on ration day would be the best moment for him to flee southward to join the ghost dancers, probably to put himself at their head. Now, five days prior to the time McLaughlin and Drum had planned to make the arrest, the agent's Indian police brought in what seemed to be certain information that Sitting Bull and his followers were packing up and preparing to leave. Thus, on December 14, Sitting Bull himself precipitated the crisis, and secret preparations were instantly made to arrest him at dawn on the following day.

McLaughlin had been stiffening and reinforcing his Indian police all through the summer in preparation for this final act, and he now had the finest and boldest force on the Sioux reservation: men who would obey any order and would shoot promptly and effectively if necessary. On December 14 he had about forty of these men hovering near Sitting Bull's camp, on the north bank of Grand River forty miles southwest of the agency. He sent a runner with orders for this force of police to enter Sitting Bull's house at dawn and make the arrest. Meantime two troops of the Eighth Cavalry from Fort Yates at the agency were to make a night march, arriving at Sitting Bull's camp in time to support the police if there should be trouble.

There was trouble, and in surprising quantities. The Indian police acted swiftly and skillfully. Before dawn on the fifteenth, twenty-nine policemen and four specials slipped quietly into Sitting Bull's camp of log houses and tipis and surrounded the chief's big log dwelling. The leading police officers went in and put Sitting Bull under arrest. He submitted quietly, asking for time to put on his clothes; but his purpose was probably to seek delay in the belief that his people would quickly come to his aid. Someone had instantly aroused the camp, and almost at once frantic yells arose and ghost dancers began to assemble outside the chief's cabin. The police hustled Sitting Bull out into the open, and without a moment's hesitation the desperate man shouted for help. He knew what that meant. A ghost dancer fired a rifle, wounding Lieutenant Bull Head. The wounded police officer whirled and shot Sitting Bull dead as a blaze of rifle fire burst from both parties.

The police stood like rocks, exchanging rapid fire at pointblank range with the howling ghost dance fanatics, who were presently forced back. The police dragged Sitting Bull's body and the dead and wounded policemen into the chief's cabin and prepared to hold out until the cavalry appeared. It was a dreadful affair, even the women among Sitting Bull's frantic followers joining in attacking the police. Inside the cabin the chief's own women armed themselves with knives and clubs and the police had much trouble in disarming them. The ghost dance warriors had retired, many to the near-by wood, from which they kept up a ceaseless fire.

At seven that morning an Indian policeman met the cavalry, commanded by Captain E. G. Fetchet and Lieutenant E. H. Crowder,[9] and informed the officers of Sitting Bull's death and the situation of the police in the camp. Advancing rapidly, the cavalry was fired on by the ghost dancers in the wood near the Sitting Bull cabin. Fetchet put his howitzer into action, shelling the wood, and as the gun opened fire a white flag was put up on Sitting Bull's roof—a signal that had been arranged to show the troops exactly where the police were stationed. Some ghost dancers gathered on a hill, and from there they started to flee southward, across Grand River. Captain Fetchet sent one troop to head them back and drive them up the river. This was to prevent their going south and leaving the reservation completely. The rest of the cavalry now advanced on foot into Sitting Bull's camp, and the Indian police came out of the chief's house and helped the troops to drive the ghost dancers out of the near-by wood. The retreating Sioux attempted to make a stand at some log cabins two miles above Sitting Bull's house, but Lieutenant Crowder attacked them vigorously and drove them higher up the river.[10]

If any prestige was to be gathered from this savage fight, the Indian police certainly won it all. With a force of 33 men they

[9] In later years a brigadier general, organizer and administrator of the army draft in World War I.

[10] *Report of the Secretary of War* (1891), I, 182; *Report of the Commissioner of Indian Affairs* (1891), 338. One Sioux version states that inside the cabin Sitting Bull's youngest son jeered at him for submitting to arrest. The chief with a policeman on each side of him had just stepped outside the door when the shooting began. Sitting Bull, seven ghost dancers, and seven Indian policemen were shot down on a patch of ground hardly twenty feet across.

had deliberately entered this hornets' nest to face over 150 fanatical ghost dancers who (thanks to the liberality of Sitting Bull's Philadelphia lady friend)[11] were well supplied with modern repeating rifles, Colt revolvers, and ammunition. Many of the ghost dance women were as dangerous as the men, slipping up on the police while they were fighting the men and striking them with knives or clubs. Most of the police losses were suffered within a minute of the opening of fire. Lieutenant Bull Head and five men were killed and one man badly wounded. The ghost dancers were reported as having eight men killed and three wounded, all on Sitting Bull's doorstep, among the dead being Sitting Bull, his seventeen-year-old son Crow Foot, his adopted half-brother Little Assiniboine, and Black Bird, a ghost dance leader.

This affair had turned out to be much more serious than anyone, with the possible exception of some of the Indian police, had thought possible, and McLaughlin must have been very thankful that he had waited for explicit orders from Washington before acting. Even so, he was nearly ruined. After all his years of service which had gained him the reputation of being the best agent the Sioux had ever had, he was now violently attacked both on the frontier and in the East. One would have supposed him the deliberate plotter of the murder of an innocent person. Some of his critics attempted to depict Sitting Bull as a mild and harmless old man, and to such persons it was clear that McLaughlin was a murderer who had planned the crime in cold blood. There can be little doubt that the officials in Washington would have thrown this agent to the wolves if he had not carefully kept the orders he had received from them, showing clearly who had directed this attempt to arrest Sitting Bull.

[11] McLaughlin's report for 1891 contains a brief account of this woman. A wealthy Philadelphia widow, she had yearnings to help the poor Sioux. She came out to Standing Rock and practically adopted Sitting Bull, who was not in the least poor and far from being an ideal house guest for a Christian lady to take in; yet she invited him to visit at a home she had established, fed him lovely meals, and loaded him with gifts, mainly money, which Sitting Bull invested in Winchester rifles, Colt revolvers, and ammunition. Agent McLaughlin was infuriated; but the woman was a prominent and highly respectable philanthropist and he could not take any stiff action against her.

12

Disaster

SITTING BULL was dead at last, but his ghost haunted the unhappy officials. He had had about 375 followers, practically all of his own group, the Hunkpapa Sioux; in the main they were veterans of the war of 1876. Of this number some 250 came to Standing Rock Agency after his death and quietly surrendered. The remainder of the group fled southward to the camp of Chief Hump on Cheyenne River, at the mouth of Cherry River. Their escape was a most unfortunate and unexpected contretemps, coming at the exact moment when at all the southern agencies the crisis seemed to be past, with all the ghost dancers on the upper Cheyenne River and in the badlands ready to move to the agencies and join the friendly camps there.

At this moment came the flying tidings that Sitting Bull had been killed and that part of his people had appeared on Cheyenne River, half-clothed, freezing, hungry, and panic-stricken, and those Sioux on Cheyenne River who saw these fugitives and heard their story leaped back at once to the old belief that the whites were not to be trusted, that the soldiers would disarm and dismount all the ghost dancers and then either kill or imprison them.

Just at this time Chief Hump had accepted the advice of his friend Captain Ewers and was preparing to take his followers from their Cherry River camp down Cheyenne River to the agency. Hump's people now became fearful of the consequences of moving to the agency where the troops would be in control. Ninety miles to the west of Hump's camp, at the forks of Cheyenne River, was the most distant of the camps of this agency, that of Chief Big Foot. This man was a son of the old Miniconjou chief, Lone Horn of the North, and he and his followers who numbered about 550 were former hostiles who had taken part in the war of 1876. They had a reputation at the agency of being troublemakers and their camp was a hotbed of ghost dancer fanaticism. Still, they were considering seriously following Chief Hump's lead and going to the agency. It was while the Sioux in Big Foot's camp were going through the slow process of making up their minds whether to move to the agency or not that the news of the fight in Sitting Bull's camp threw them into intense excitement and doubt. They were angry at the killing of their fellow tribesmen in the Sitting Bull camp and they feared that their own camp might be similarly attacked if they moved toward the agency and came closer to the troops. While they were hesitating as to what course to follow, Colonel E. V. Sumner with a force of cavalry came down on them from an unexpected direction, moving down Cheyenne River from the Black Hills district, and taking them off guard he put the whole band under arrest. This was on December 22, one week after the killing of Sitting Bull. Colonel Sumner, who had had much experience among the Sioux, was favorably impressed by Big Foot's conduct. The chief seemed friendly; he said that he had just received an order to bring his band to the agency and he was on the point of starting. After careful consideration Sumner decided to permit the Indians to go down to the agency without a guard of troops, and Big Foot said good-bye and went his way.

All now seemed to be going well. The death of Sitting Bull had ended all danger at Standing Rock; at Cheyenne River the last of the ghost dance bands, Big Foot's, was on its way to the agency. Lower Brûlé was quiet.[1] Rosebud was peaceful; at Pine Ridge a great mass of friendlies was encamped at the agency under control of the troops and the ghost dance camp in the badlands, on the plateau about fifty miles north of the agency, was being constantly weakened by desertions, one group after another

[1] At Lower Brûlé some of the Sioux had started ghost dancing, but the agent promptly used his police, broke up the dances, and sent seventeen of the leaders to Fort Snelling in Minnesota. That ended the trouble. The tame Sioux at Crow Creek agency did not take up the ghost dance craze.

slipping away to go to the agency. From his headquarters in the Black Hills General Miles was directing his troops, who had the Indians surrounded and were in a position to move in on them from any direction. Miles had over half of the United States Army drawn around the Sioux reservation and he could have delivered a crushing blow at the ghost dancers at this moment; but he had high hopes of ending the trouble without further bloodshed. The anger and fear caused by the death of Sitting Bull and his followers was the only deterrent to a peaceful conclusion.

The news of the killing in Sitting Bull's camp spread alarm through the friendly camps at Pine Ridge. On December 22, a force of cavalry came on a party of Sioux fleeing from the friendly camp to the ghost dance stronghold in the badlands. The Sioux were unable to travel rapidly as they were driving a large herd of ponies before them, and after an exchange of shots the cavalry turned the Indians and drove them back to the agency. This created much excitement, and that same night the Star Pansy messiah from Iowa turned up in Red Cloud's camp at the agency, almost precipitating a fight. The following morning, in a sudden panic, the Brûlés who had been induced to leave the ghost dancers and camp at the agency, fled back toward the badlands, taking Chief Two Strike and Kicking Bear, the Cheyenne River ghost dance leader, with them. Kicking Bear, a real fanatic, was bent on mischief. As soon as he reached the ghost dance camp he organized a war party and headed westward toward the white settlements in the Black Hills. On Battle Creek he ran into a small force of Cheyenne Indians who were serving as scouts with the troops and attacked them; but after a hard fight in which several Indians were killed the ghost dancers were forced back into the badlands.

A scouting party sent out from the Black Hills by Colonel E. A. Carr of the Sixth Cavalry now discovered the ghost dance camp on the badlands plateau, on the trail running from the mouth of Rapid Creek eastward and southward to the mouth of Wounded Knee Creek on the Pine Ridge reservation. This camp consisted of seventy lodges, not more than seven hundred people, and these were the only Sioux from whom any act of hostility was to be expected. The only outlet westward through which the ghost dancers could hope to escape was by a trail through Cottonwood Canyon, and Carr now placed a force in the canyon, bottling the Sioux up. This precaution seemed hardly necessary; for the friendly Sioux from the camps at Pine Ridge Agency had been plying the ghost dancers with the strongest arguments against attempting to hold out any longer, and on December 27 the ghost dance leaders decided to give up. They broke camp and began a slow movement southward toward Pine Ridge Agency. In their rear a screen of troops followed, to persuade them not to alter their decision, the soldiers so close on the heels of the Sioux that they often came to campfires still burning, where the Indians had broken camp to move on toward the agency. The ghost dancers hesitated, then took the plunge and crossed White River, camping on Big Grass Creek. Their next move was to Big White Clay Creek, where they encamped at or near the deserted log cabins belonging to Man-Afraid-of-His-Horse's Payabya Band. From here they visited the Roman Catholic mission, only four miles from Pine Ridge Agency, to find out if their friend Father Jutz really thought it safe for them to go on to the agency. At this moment, when the final act of the ghost dance troubles was seemingly about to end, the vindictive ghost of Sitting Bull whirled onto the stage, and instead of peace came catastrophe.

It was with Big Foot on Cheyenne River that Sitting Bull's spirit came to dwell. Big Foot's permanent camp with many log cabins was on the south bank of Cheyenne River at the mouth of Deep Creek. It was the most distant camp on the Cheyenne River reservation. Sumner with his cavalry was in position west of this camp and, as we have seen, had obtained from Big Foot a promise to move his band eastward to the agency. On December 15, the day of Sitting Bull's death far away to the north on Grand River, Big Foot came to Sumner's camp to state that he was taking his band to the agency for annuities: the annual winter distribution of clothing, bolts of cloth, canvas for new tipis, kettles, knives, and other necessaries. So Big Foot departed in peace; but a day later, far down Cheyenne River, news of Sitting Bull's death reached his camp, and at about the same time some of his young men who had gone on ahead reported that infantry was advancing up Cheyenne River. This was the force under Colonel H. C. Merriam, whose movement from Cheyenne River Agency westward had been ordered by telegraph by General Ruger, who was in St. Paul, out of touch with the situation. Merriam was doubtful of the wisdom of such a march. He was afraid that the Big Foot people would regard his movement as hostile and would stampede in a panic.

Big Foot halted his movement toward the agency. Like Two Strike of the Brûlés, this Miniconjou Sioux was a sturdy, honest old fighting chief who found thinking a slow and difficult business, and here he was faced with a complicated situation fitted to tax the nimblest of wits. Sitting Bull was dead in his camp up on Grand River, and by the very best Sioux reports that chief had been the victim of treachery and plotting hatched by the agent and the white soldier-chiefs. Here was a force of infantry advancing up the river toward Big Foot's camp; there was Colonel Sumner up the river, supposedly his good friend, but now behaving in a very suspicious manner, sending cavalry to dog his steps, as if preparing for an attack. Indeed, on news of the fight in Sitting Bull's camp, General Miles had become alarmed over Sumner's permitting Big Foot to move toward the agency without a guard and had sent peremptory orders for the cavalry to follow Big Foot, round his band up, and hold it.

With Big Foot bewildered by his situation and some of the ghost dance fanatics in his camp shouting defiance and preparing themselves to fight, a party of fugitives from Sitting Bull's camp made their appearance and threw Big Foot's people into intense excitement. There were about thirty-eight of these Hunkpapas from Grand River: men, women, and children. They were in a deplorable condition—starving, all suffering from lack of clothing, some with frozen hands and feet. The sight of these unfortunate kinsmen and the terrible story they told filled the hearts of Big Foot's followers with pity. Many it enraged. With Big Foot puzzled as to what to do and with his people excited, fearful, angry, and quarreling among themselves, the band halted in the log cabin camp which Chief Hump's people had abandoned when they had set out for the agency a few days previously. Here Big

Foot called a council, but his leading men were widely divided in their opinions as to what should be done.

The Sioux always said that when white soldiers appeared big trouble was sure to follow, and the course of events on Cheyenne River seemed to bear this out. Everything went wrong. Colonel Sumner, ordered by General Miles to pursue Big Foot and take his band into custody, replied that he considered such a course unnecessary, in that Big Foot was friendly and was already moving down the river on his way to the agency. Sumner had hardly sent off this message when he received word that Big Foot had been joined by some of Sitting Bull's people and had instantly halted his movement toward the agency. Sumner now made a rapid advance toward Hump's camp, where the Big Foot band was halted. Both sides were now very suspicious; but on the twentieth, Big Foot came to Sumner's camp as friendly as ever, to make excuses for taking in Sitting Bull's people. He said that they were relatives of his own people and that they would die of starvation and cold if not helped. Sumner, who had heard of the condition of these fugitives, told the chief that he was quite right to succor them. Still, unwilling to take chances, Sumner now sent an officer with Big Foot, to spend the night in his camp and to bring the whole band back up the river the following morning.

On December 21, Big Foot and his band came quietly back up the river to meet the cavalry. The chief had told Sumner that his band consisted of 100 of his own people and 38 of Sitting Bull's folk; but now on counting them, Sumner found that there were 333 and his suspicions concerning Big Foot's good intentions became aroused again, although a bad interpreter may have been to blame for the discrepancy in the number given by the chief. This matter hardened the Colonel in his determination to force the band to return with his troops up the river; but the march had hardly started when some young Indians attempted to ride on ahead and were ordered back, the troops lining up menacingly to enforce the order, and instantly the women and children began to scream and run. Big Foot quieted them, but it was apparent to everyone that most of the Sioux were panicky, while the rest—

the ghost dance fanatics—were defiant. When the marching column at length reached Big Foot's old settlement on Deep Creek, each family fairly scuttled into its own log cabin and barricaded the door. Big Foot told Colonel Sumner mournfully that it would be very difficult to get his people to come out. They were frightened, and mad. Sumner's own view was that any attempt to get them out of the cabins would probably bring on a fight. A humane man, and still trusting the chief, he told Big Foot that he would withdraw his troops to a camp farther up the river, trusting him to keep his people quiet.

At this moment Sumner's attention was distracted by a message from General Miles, reporting that a large force of Sioux was coming down from the north toward Cheyenne River and ordering Sumner to head them off. The report was a false one; but before Sumner found that out, Big Foot's people, with or without the consent of their chief, left their camp on Cheyenne River and fled southward into the Big Badlands. Sumner was very angry, and rightly so. He had exceeded his orders to humor these Indians; he had trusted their word, and they had betrayed him; yet there can be little doubt that Big Foot had acted in good faith and that his people had broken their promise and fled because they were in a panic. In heading south into the badlands, they had two choices open to them—the ghost dance camp and war, or Pine Ridge Agency and peace—and although a party of sixty-five, including eighteen warriors, did leave the main body and go off to seek the ghost dance camp, the trail of Big Foot's main body, which was traced by the troops in pursuit, passed far to the east of the plateau where the ghost dance camp was located and struck straight on, crossing White River near the mouth of Eagle Nest Creek and then turning straight for Pine Ridge Agency.[2]

On learning that Big Foot's band had eluded Colonel Sumner, General Miles telegraphed to General Brooke at Pine Ridge, ordering him to send out troops to apprehend the fugitives. Brooke dispatched Colonel Guy V. Henry with four troops of the Ninth

2 Big Foot Pass, marking the point where that chief's band descended from the badlands to White River, is midway between the present towns of Interior and Quinn.

Cavalry and three Hotchkiss guns, carried on muleback. Henry combed the country north and northeast of Pine Ridge, but could find no Indians. General Brooke now sent out Major S. M. Whitside on a similar mission with four troops of Custer's old regiment, the Seventh Cavalry, who marched twenty miles toward the northeast through a terrific sand storm that blinded and almost suffocated the men. The great drought of 1890 was still holding on; there had been no snow to speak of at Pine Ridge for weeks, and when the bitter gales swept in from the northwest howling across the empty and desolate country, flying clouds of sand and gravel made traveling almost impossible. Near a small deserted Indian store on upper Wounded Knee Creek, Whitside's troops discovered Indians in scattered groups and started a pursuit, but the Sioux got away. The cavalry camped on Wounded Knee for the night, and on the morning of the twenty-eighth Major Whitside sent out a scouting party which was accompanied by the famous Sioux mixed-blood scout, Little Bat. Ten miles east of camp, on Porcupine Tail Creek, the scouts came on a portion of Big Foot's band. With troops out in every direction combing the country for them, these Indians in scattered groups were moving across the prairie toward Pine Ridge Agency as quietly as lambs. Major Whitside, informed by his scouts that they had found the quarry, rushed his troops into concealment behind a ridge, and as the straggling groups of Big Foot's band drew near, he brought his troops out into the open and sent his Sioux scouts to demand that Big Foot's people should surrender. After some parleying with the Sioux scouts, Big Foot gave his assent, and his people, surrounded by the troops, were marched to Wounded Knee, where they put up their tipis close to the tents of the cavalry. The point where the troops and Indians were encamped was on the west bank of Wounded Knee Creek, about eighteen miles northeast of Pine Ridge Agency. It was the locality that was known in the early 1880's as the Red Dog camp, whose leading man was old Chief Red Dog, a friend and ally of Red Cloud. Now, in December, 1890, the scattered log cabins and little day school (Number Seven) lay deserted, part of the Sioux being in the friendly camp at the agency, the rest with the ghost dancers.

Major Whitside sent a message to General Brooke at the agency reporting that Big Foot's band had surrendered quietly to his command, and Brooke immediately dispatched Colonel George A. Forsyth with another battalion of the Seventh Cavalry to reinforce the troops at Wounded Knee. Forsyth got to Whitside's camp during the night and took over command as senior officer. He now had in camp eight troops of cavalry, 470 officers and men, with a group of Sioux Indian scouts and four pieces of light artillery. Big Foot had told Colonel Sumner that he had 100 people plus 38 Sitting Bull fugitives in his camp on Cheyenne River. Sumner had then counted and found 333. Now on Wounded Knee another count showed 370, and this did not include a party of 65 who had gone off to seek the ghost dancer camp in the badlands. This unaccountable fluctuation in the number of Sioux in any given camp from day to day was why census-takers on the reservation went mad. The shift in figures worried the military commanders; but in this instance Major Whitside had taken a very careful count of males and had found 120 men and boys fit to fight, if they were mad enough to try fighting a much better armed force of 500 troops and scouts.

Big Foot did not wish to fight. The old man had pneumonia, and how he had kept in the saddle and continued his march through the bitter cold, only a Sioux of his day could explain. He had been very glad to give up, and now he was in a tent in the soldiers' camp, a big Sibley tent with a hot stove in it. Most of his followers were evidently also glad that their flight was ended and that they were to have peace and security; but some of the ghost dance fanatics were smoldering with suppressed rage and watching the soldiers like hawks, waiting for the moment when they might find the opportunity to strike. Colonel Forsyth's orders were to disarm these Indians and march them to the railroad at Gordon, Nebraska, where they would be put on a train and taken to a military post, to be held until the trouble on the reservation was ended. The order was a terrible mistake. These Indians supposed that they were to be permitted to join the other Sioux in the camps at Pine Ridge Agency. Once among their own people and quieted down a little they would have given up their arms without the least trouble; but they were now very nervous and suspicious, and to demand weapons from Sioux in that state of mind was inviting tragedy. Colonel Forsyth was a veteran Indian fighter and certainly knew all this. The very preparations he made show that he knew. If Colonel Sumner had been in command, he might have ignored his orders and humored Big Foot's people for the sake of peace; but Forsyth was the Custer type of officer and simply did not understand why he should humor Indians to avoid a fight. Sumner had humored these Sioux, and they had broken their word to him and run away. Colonel Forsyth had no intention of negotiating with these runaways. He would issue his orders to them and see to it that they obeyed.

Early on the morning of the twenty-ninth he drew up his troops, completely surrounding the Indian camp, with the Hotchkiss guns on high ground trained on the camp and ready to open rapid fire with explosive shells. Indian interpreters went into the camp and ordered the men to come out and form a line. In the cavalry camp the officers had formed a group in front of the tent which sheltered the sick chief, Big Foot. Colonel Forsyth had a talk with the chief, Philip F. Wells acting as interpreter. Big Foot stated that his men had no arms, which was a palpable lie and a plain indication that neither the chief nor his followers trusted the troops or were willing to give up their weapons. The Sioux men and boys were now moved out of the Indian camp. They slouched forward and squatted on the cold ground in a semicircle in front of the tent in which their chief sat. The men and boys were all wrapped closely in their blankets, under which many of them wore ghost shirts which the fanatics had assured them no bullet could penetrate. That most of these Sioux believed this we cannot doubt. Facing the squatting Sioux and very close to them was a line of dismounted cavalry, the men alert and all holding their carbines at the ready.

Colonel Forsyth now issued an order for the Indian men to return to their camp immediately behind the spot where they were sitting, twenty at a time, and to bring out all their weapons. The interpreter explained this, and the first twenty slouched back among the tipis. How well one knows this ancient ritual, which

Courtesy Bureau of American Ethnology

Indian police at Standing Rock after fight in Sitting Bull's camp.

Courtesy National Archives

Great Sioux camp at Pine Ridge, 1891.

the army had tried on wild Indians so many, many times in the past! The ritual was always the same. The first twenty Sioux came back from the tipis with two ancient muskets. Colonel Forsyth and Major Whitside conferred, an order was given, and a group of cavalry started for the Indian camp to make their own search. The Sioux women and children, huddled together in front of the tipis, were clearly very much frightened, and some of the women now began to wail. Major Whitside, in direct command of the troops in line facing the Indian men, now ordered his men to take several paces forward; and as the soldiers came on, the squatting Sioux lost their air of stolid indifference and watched every movement of the whites with keen-eyed suspicion. The troopers who had entered the camp came out with about forty guns, mostly old muzzle-loaders, and a few revolvers. Colonel Forsyth was clearly annoyed at this cache of almost useless arms. He started to speak to Major Whitside; but now an Indian medicine man, Yellow Bird, jumped up from the line of squatting Sioux and began to trot around in a circle, blowing on an eagle-bone war whistle and stooping every now and again to snatch a handful of dust and throw it into the air. This was ghost dance magic. Perhaps it symbolized the beginning of the great dust storm that the ghost dancers believed would bury the entire white race. This fanatic presently stopped his trotting and began to harangue the Sioux in a high excited voice. The Indian scouts understood his words and jumped to the alert, some of them shouting in Sioux for the squatting warriors to stay where they were and not to make a move. Many of these warriors were now singing their death songs —a very bad sign.

Colonel Forsyth curtly ordered Philip Wells, a scout and interpreter, to tell the medicine man to stop his harangue and sit down. Yellow Bird started to obey the order; but at the same instant a Sioux pulled a gun from under his blanket and fired, wounding an officer. As at a signal, the warriors leapt to their feet, throwing off their blankets, and at the same moment the alert troops poured a volley into them. On the high ground close to the camp Captain Capron saw the first shots fired and at once ordered his Hotchkiss guns into action. They rained explosive shells among the Sioux women and children in the Indian camp. Everyone knew that no Sioux men were in that camp.

Philip F. Wells was standing among the officers close to the Sioux men when the fight started. He was of part Sioux blood and had spent most of his life with the tribe, speaking Sioux perfectly and knowing Sioux ways intimately. He stated in later years that he saw a Sioux fire the first shot, and that from what he heard and witnessed he believed the Indians had a plan to break through the screen of troops and run to some small canyons along Wounded Knee Creek. To do this they had to break the line of Troop "K," and despite the fact that they could not fire on that troop without killing their own women and children who were gathered between the troop and the Indian camp, they did open fire and some of the men and boys actually broke through and escaped up the creek canyons. But most of the Sioux males died under the murderous fire of the carbines on the spot where the fighting began, their bodies lying in a tangled semicircle just where they had squatted during the search for weapons. Philip F. Wells stated that the fight did not last more than two minutes, and then the Sioux men, women, and children were fleeing in every direction with the troops in hot pursuit. It was during this chase that most of the women and children were shot down. Some of the officers shouted not to kill women; but the men had the usual excuse that Sioux women were difficult to tell from men and that some of them had weapons. There were plenty of Sioux mixed-blood scouts with the troops to point out the men; but in the heat of the pursuit most of the soldiers made no attempt to distinguish between either sexes or ages. The Sioux scouts saved many of the women and children; but while they were doing all they could the troops were hunting down the fugitives in the canyons and little ravines and shooting indiscriminately into the mixed groups of Indians. The exact losses of the Sioux in this bloody affair have been stated variously, but a careful count made in 1891 shows 84 men and boys of fighting age, 44 women, and 18 children killed. There were at least 33 Indians wounded, many of them mortally. Of 106 men and boys who had squatted in the semicircle closely surrounded by dismounted cavalry, 52 died on the spot; most of the others died in hand-to-hand fighting while trying to break away or after reaching the canyon of Wounded Knee or one of the little ravines.

Eighteen miles away across the prairie at Pine Ridge it was a fine crisp and sunny December morning. Some of the Sioux knew that the troops had rounded up the Big Foot band, and they were excited about that; but in the friendly camps to the west and south of the agency buildings and in the ghost dance camp one and one-quarter miles north of the agency all was quiet. The ghost dancers, moving in with timid caution, first from the badlands, then venturing across White River to the vicinity of the Roman Catholic mission, had now taken the final step and camped close to the agency. The time of fear seemed ended, and on this fine morning friendly smoke was drifting into the sky from a thousand Sioux tipis and from the near-by Sibley tents of the white soldiers. But now a dull throbbing sound disturbed the still morning, coming from far away to eastward—the sound of Capron's guns pouring explosive shells into Big Foot's camp, and the Sioux in the Pine Ridge camps rushed out of tipis to listen. Parties of mounted men assembled and shot out of the camps; the ghost dance camp north of the agency became the scene of wild confusion, some of the families hastily taking down their tipis and preparing to flee, while Two Strike and some of the other leaders tried to quiet the fears of their panicky followers. Bands of mounted Sioux streaked off toward Wounded Knee, and in an incredibly short time mounted runners came flying back to the agency with the news that the white soldiers were killing Big Foot's people. At this the whole mass of Indians encamped around the agency went mad, some with rage and others with fright. Tipis went down in the camps and family groups made off toward the hills; bands of armed warriors drew together and began to argue loudly as to what should be done. A few bad boys from the ghost dance camp slipped up close to the agency and fired some shots. They also seemed to be trying to set fire to some buildings, and Agent Royer put all fifty of his policemen in line, to fight them off. The agent, as they tell it at Pine Ridge today, then sought a safe cellar, and presently General Brooke ordered the police to stop shooting. "You pop at one of those fellows and ten of his friends come running to help him." That was true; and with all his cavalry away Brooke did not have the men to defend the agency from a real attack. He let the ghost dancers amuse themselves until they got uncomfortably close, then sent some troops to aid the Indian police, and they drove the ghost dancers back across the hill where the government hospital now stands.

After midday the Seventh Cavalry troops began their march back to the agency, leaving the field of Wounded Knee in groups. The Sioux from the ghost dance camp (Brûlés) ran into a column commanded by Captain Jackson and instantly assaulted it, forcing the troops to drop twenty-three Sioux women and children prisoners. The Sioux now fired the prairie, and as twilight came on, the Seventh Cavalry groups were making their way through dense smoke and gathering darkness, with parties of Sioux appearing out of the gloom, firing some shots, and vanishing again. Thus with the Indians hanging on their front, flanks, and rear, the troops marched into Pine Ridge, bringing in their twenty-five dead and about thirty-five wounded troopers. They also had a few Sioux women and children with them, some badly wounded.

The Sioux from the agency had picked up some of the survivors of Big Foot's band, in the main women and children and wounded; when they heard the story of these unfortunate kinsmen, they went mad with rage. Runners on swift ponies shot away westward, to take the word to the agency camps that Big Foot's band had been first disarmed and then deliberately killed.[3] When the runners spread this report among the camps at the agency and added that the troops from Wounded Knee were marching back to Pine Ridge, the Sioux went wild. The ghost dancers had already retired toward their stronghold in the badlands; now the people in the friendly camps began to take down their tipis and load their belongings on ponies or into farm wagons. As one group after another headed north toward the badlands, the panic spread through the camps, and more and more of the Indians were affected. Many

[3] This first report of the fight the Sioux heard was communicated by one of the Pine Ridge Indian policemen in a letter to his sister at Cheyenne River Agency. On January 11, 1891, Colonel Meriam reported that this letter with its tale of the massacre of Big Foot's unarmed people had almost caused an outbreak among the Cheyenne River Sioux.

of the younger men declared that they would stay and fight the returning cavalry; but the men with families thought only of getting away. The valley of Big White Clay Creek was soon black with hurrying groups of Indians; and in the midst of this panic-stricken throng old Red Cloud was being borne along on the last big adventure of his exciting career. That morning there had been some six thousand Sioux in the peaceful camps around Pine Ridge Agency. By dark, four thousand had fled, leaving at the agency about two thousand badly frightened Sioux who had remained, some of them because they still were trying to trust the whites, others because they had no means of transportation and could not get away[4]

With the consent of the Reverend Mr. Cook (a missionary of Sioux blood) the pews were taken out of the handsome little Pine Ridge Episcopal chapel and the wounded Indian women and children from the battlefield were laid on the floor. In the midst of this scene of horror stood a Christmas tree, beautifully decorated. All through that bitter night, under the lamps and candles, young Dr. Charles A. Eastman, the Sioux mixed blood, labored anxiously to save the wounded women and children of his own tribe. The previous August, filled with fervor, he had told the members of the Indian Friend groups at their annual Lake Mohonk conference how he had prayed and striven for many years to escape from Indian barbarism into civilization. His thoughts on civilization must have been peculiar as he bent anxiously over the dying Indian women and children in the little chapel. McGillycuddy, who had been an army surgeon before he was made agent at Pine Ridge, was also in the chapel, doing all he could to save the wounded. It was known that a large number of injured women and children still lay on the field of Wounded Knee, but nothing could be done at present to succor them. The country between the agency and Wounded Knee was

[4] Among the best printed accounts of the Wounded Knee disaster are those of General Colby in *Nebraska Historical Society Transactions*, III; W. F. Kelley in *Nebraska Historical Society Transactions*, IV; Philip F. Wells in *North Dakota History*, October, 1948; Charles A. Eastman, *From the Deep Woods to Civilization*. The officials reports are in the War and Indian Office volumes for 1891.

swarming with mounted parties of enraged Sioux who were ready to attack any group of whites they might meet.

Slowly that endless night wore on. No one tried to sleep. The Negroes of the Ninth Cavalry were in the badlands ninety miles away to the north when a courier reached them with orders to come down at top speed to the relief of the agency. They marched southward all through the night, getting in at dawn on the thirtieth; but they had hardly dismounted at the agency after their amazing ride when word came that their pack train laboring in their rear was being attacked by the Sioux at a point two miles north of Pine Ridge. They mounted at once, marched to the scene of trouble, and drove the Indians off. As day came, smoke was seen billowing up to the north of the agency, and it was reported that the ghost dancers had fired the Roman Catholic mission. Colonel Forsyth now took eight troops of the Seventh Cavalry and marched down Big White Clay toward the mission, taking one Hotchkiss gun with the force. He found that the mission was not under attack; but a large force of Sioux led by Two Strike and Little Wound had set fire to the empty log cabins of the White Bird (Spleen Band) camp. A government day school was also going up in smoke. Following hotly a group of retreating Indians (which his long experience should have warned him was a decoy party), Colonel Forsyth found his command trapped on low ground with the Sioux in strong force holding all the surrounding hills. Thus in their last brush with the United States Army the Sioux employed with success this ancient and simple trap which had always been about the only thing they had in the way of strategy. The situation of the Seventh Cavalry was bad; but now Colonel Guy V. Henry and his Negro cavalry suddenly appeared on high ground, their trumpets sounding as they bore down on the encircling Sioux and swept them aside, releasing the crestfallen Seventh Regiment from the tight place it had blundered into. The Brûlés and Oglalas retired to their camp seven miles north of the agency, and Forsyth and Henry, deciding that the Indians were in too strong force to justify an attack, drew off to their camps at Pine Ridge.

On this day, December 30, the military were much disturbed

over the situation at Pine Ridge. The Indians, still raging over the killing of Big Foot's band, were very active, swarming over the country in mounted bands and apparently hunting for a fight. General Miles was hastening to Pine Ridge with cavalry reinforcements. The weather had worsened and a dry blizzard was raging: a terrible cold wind, but almost no snow. Despite these conditions, a party of civilians and friendly Indians left the agency and went to the Wounded Knee field to search for wounded Sioux who were known to be lying there in the bitter cold, covered by a thin blanket of snow. They found dead women three miles from the point where the fight had started and other bodies dotted along the way to Big Foot's camp. The Sioux who were with the party kept up a dreadful wailing. They found the Indian camp burnt and a heap of men's bodies where the fight had started. A few wounded women and children were still alive, but the Indian men were all dead. Ghost dancers were seen, riding up and down on tops of near-by buttes, and a messenger was sent to the agency to ask for an escort of troops. This party returned safely, bringing with them the few Indian women and children they found alive.

On this day a party of sixty-five Indians from Big Foot's camp, including eighteen men, were brought safely to Pine Ridge by the Sioux scouts. This was due to the action of Philip F. Wells, who was in charge of the Sioux scouts; for on the morning of the Wounded Knee fight he obtained a hint that some people had left Big Foot's camp before the troops had discovered it, and he sent Sergeant Standing Soldier with ten men to scout eastward through the Corn Creek and Bear-Runs-through-Lodge Creek district. Here they found the group from the Big Foot camp. These people were fugitives from Sitting Bull's camp who had apparently decided that it would not be safe for them to go to Pine Ridge with Big Foot. The Sioux scouts took them to Pine Ridge by a roundabout way, southward into Nebraska, and came into the agency without running into any of the ghost dance parties.

Bad as the situation seemed on December 30, it began to clear up in the first days of the new year. General Miles arrived with fresh cavalry forces and took personal charge. He relieved Colonel Forsyth and put him under arrest because of the reports of the killing of women and children in the Wounded Knee fight. Young-Man-Afraid-of-His-Horse now came home from his hunting trip in Wyoming and vigorously espoused the cause of peace. For a time he and American Horse were the only important chiefs in the friendly camp at the agency. The other important men—Red Cloud, his son Jack, Little Wound, Big Road, Two Strike, Crow Dog, and several others—had been swept away in the panic flight from the agency camps when the first news of Wounded Knee was brought in. These chiefs had been compelled to accompany the mass of frenzied Sioux; but they had now quieted down and were working to separate their people from the ghost dancers. That was not an easy task; for the ghost dance fanatics, making full use of the dreadful story of the fate of Big Foot's band, were telling the Sioux that the whites could not be trusted and that the only thing to do was to stay where they were and fight. This was a strong argument; but there was a still stronger one working for peace. These Sioux from the agency camps were already missing the regular issues of beef and other rations. They were hungry and cold, and most of them were not interested in fighting the troops. At this time Red Cloud was reported to have sent a message to the agency, begging that troops should be sent to rescue him from the ghost dance camp. From this final shame of having to be saved from his own people by white troops the old man was delivered by his courageous daughter who led him, blind and helpless, out of the ghost dance camp into a howling blizzard and brought him safely to his own camp at the agency.

Young-Man-Afraid-of-His-Horse had already contacted the Sioux in the camps north of the agency, and he was advising them to trust General Miles, to make the best terms they could with him, and then to move in to the agency and surrender. The General was emphasizing the wisdom of this advice by slowly moving strong forces of troops down from the north, gently pressing the Sioux away from their only stronghold in the Big Badlands. Other columns were moving in from the west and east, and at Pine Ridge Agency there was a sufficient force to defeat any attempt of the

Indians to attack there. True, the Sioux might have made a desperate plunge toward the southeast, carrying war into Nebraska; but the Indians in 1890 were like animals who had been confined in a cage for so long that they were afraid to leave, even with the door standing open. They do not appear to have even discussed the project of moving off the reservation.

Seeking new pasturage for their vast pony herds, the ghost dancers had now moved from the mouth of Big White Clay to Big Grass Creek. General Miles and Young-Man-Afraid-of-His-Horse were constantly sending friendly Indians from the agency to assure the Sioux that if they gave up they would be treated kindly. These emissaries told the Sioux that Miles now had 141 companies of cavalry and infantry—8,000 soldiers—drawn up on every side of their camp; they invited the Sioux to count their own force (600 to 1,000 men) and think it over. The chiefs thought it over and began a slow and cautious movement toward the agency, a continuous and violent quarrel going on in every camp, for most of the people were still afraid to trust the whites, and some men were loudly proclaiming that they intended to fight. Indian scouts from the agency, watching from the hilltops, could hear the uproar in these Sioux camps; at night they saw the flash of Winchester fire. That would go on all through the night; in the morning part of the tipis in the camp would come down and the families start slowly toward the agency. Next the angry families left behind would begin to feel lonesome and exposed, and they would take down their tipis and hurry after the others. Catching up with them, the whole lot would encamp and resume quarreling. To add to the turmoil, every time the Sioux moved camp a little nearer to the agency they would discover that troops coming down from the north had also moved a little farther south and were practically treading on the heels of the Sioux moccasins. Then there would be a terrific outburst in the Indian camps. Some bold spirits would loudly demand that all the men should come with them and shoot up the soldier camp; but in the end the Sioux, who could now only move in one direction, would take down their tipis and inch cautiously along toward the agency. Such was the character of the last march of the Sioux as a free people. Every mile was bringing them closer to the inevitable hour of surrender; and how they hated the thought of that!

Presently they came to the vicinity of Man-Afraid-of-His-Horse's deserted log cabin camp (Payabya Band) near the Roman Catholic mission, and here they again set up their tipis. Lieutenant E. W. Casey was at this time commanding a troop of Cheyenne Indian scouts and was taking part in the operation of slowly pressing the Sioux south toward the agency. On the eighth of January, 1891, this young officer decided to ride up Big White Clay Creek to attempt to pick up information. He rode quite close to the ghost dance camp, and meeting Pete Richards (a Sioux mixed blood, married to one of Red Cloud's daughters) and two full-bloods he got into talk with Richards and was questioning him as to conditions in the Indian camp, when a young Brûlé named Plenty of Horses slipped around behind him and shot him dead.

At the moment when all the white soldiers and civilians were exclaiming in horror over this cold-blooded murder of young Casey, news reached Pine Ridge of the brutal killing of an old and harmless Sioux named Few Tails. This Indian had obtained a pass from General Miles, permitting his family and some other friendly Sioux to go on a hunting trip north of the Black Hills. The old man was returning to Pine Ridge with his family in a wagon when one night a group of white men came to the camp. They seemed very friendly, cracking jokes with Few Tails and laughing; but in the morning they set an ambush at a creek crossing, opened fire from their concealment and mortally wounded Few Tails, also wounding his wife, who held a baby in her arms. They pursued the wounded woman, to finish their work, but she got away from them; then after plundering the Sioux belongings that lay scattered on the ground, the white men withdrew. In bitter cold weather the wounded woman wandered for seventy miles, without food or shelter, carrying her baby, until she reached a camp of a detachment of the Sixth Cavalry and was saved. Her story was confirmed by an examination of the scene of the ambush, and the murderers were found and taken into custody.

The Sioux seem to have had the better of this exchange of

amenities. After all, the white men who had shot Few Tails committed the crime for profit, while Lieutenant Casey was killed while engaged in scouting near a hostile camp and, as the jury later decided, acting in effect as a spy. Plenty of Horses, who shot Casey, had an almost ethical excuse. He said, when brought into court, that when he was a boy the agents of Captain Pratt practically kidnapped him, taking him far away to the Carlisle School in Pennsylvania and keeping him there for years while they tried to make him into a white man. They had then brought him back to the reservation, where his own people despised him because he was too much like a white man, while the whites despised him because despite all his grooming at Carlisle he was still most obviously not a white. He was neither one thing nor the other, and he was very miserable. Being drawn into the ghost dance camp and coming in contact with Lieutenant Casey, it had suddenly entered his mind that here was the opportunity to show his own people that he was as good a Sioux as any of them. He therefore shot the officer dead, in the best old-time Sioux fashion. Having obtained peace in his own soul by this curious method, he sat quietly in the court room, caring nothing, while the whites tried him for murder. The jury acquitted him, and another jury acquitted the killers of old Few Tails. In those days one had to have a real mountain of evidence to induce a Dakota jury to convict anyone of murder.

General Miles was laboring ceaselessly to restore the confidence of the Sioux in the good faith and friendly intentions of the government. As his assistants he had assembled a group of army officers whom the Sioux had known and trusted for many years. He was demanding that he be given complete control of the agencies, not as General Sheridan had demanded it in 1876 in a spirit of vindictiveness and with the object of punishing friendly Indians, but simply to insure that his generous policy should be promptly put into operation. In mid-January on authority from the Department of the Interior he replaced Agent Royer of Pine Ridge by appointing Captain F. E. Pierce, First Infantry, as agent. At the same time he appointed Captain Jesse M. Lee, who had been Spotted Tail's agent in 1877, as agent for the Brûlés of Rosebud, while Captain J. H. Hurst was made agent at Cheyenne River. Captain William E. Dougherty, First Infantry, who had been agent at Crow Creek and Lower Brûlé in 1878–80, was also on duty at Pine Ridge, and Captain E. P. Ewers, who had already performed invaluable services by inducing his old friend Chief Hump to surrender at Cheyenne River, was now brought to Pine Ridge to exert his influence among the Northern Cheyennes, who regarded him as their friend. These Cheyennes had lived in a settlement north of the agency on Big White Clay Creek; many of them had fled to the ghost dance camp, and Captain Ewers was employed to coax them to return to the agency. Young-Man-Afraid-of-His-Horse was now going boldly into the ghost dance camp, to urge the chiefs to give up. Those chiefs—Little Wound, Two Strike, Big Road, and some others—were all strongly inclined to accept his advice; but the ghost dance leaders, Short Bull and Kicking Bear, who were not chiefs, still threatened to shoot anyone who tried to leave their camp. They were now however more frightened than brave, and when Man-Afraid-of-His-Horse told them grimly just what would happen to them personally if they started shooting their own tribesmen, they quieted down perceptibly. The fact was that few of the ghost dancers had come through the Wounded Knee affair with their faith unshaken. Big Foot's men had nearly all worn ghost shirts which the prophets had assured them no bullet could penetrate; and Short Bull and the other ghost dance fanatics had seen all of these men lying dead in the snow, their bullet-pierced ghost shirts soaked in blood. That had been a shock to the stoutest believers in the messiah and his teachings.

On the evening of January 9, Young-Man-Afraid-of-His-Horse again went to the ghost camp to urge Little Wound, Big Road, and Two Strike to bring their people to the agency. All through the following day and night the Sioux quarreled, Short Bull and Kicking Bear making one last effort to prevent the surrender, but on the eleventh they broke camp and moved nearer to Pine Ridge. Here they went into camp for another grand quarrel, which this time ended in a real split, all except the hardiest of the ghost dancers leaving the camp and moving to a new location, only three miles from Pine Ridge. From this new camp Little Wound, Big Road, and Two Strike sent word to General Miles that they wanted peace and were coming in; but in their camp pandemonium reigned, the young Brûlés in particular being frantic with rage, threatening with their Winchesters any man who favored moving to the agency. On the thirteenth, part of the people left this camp and moved closer to the agency, the rest staying where they were and making a great show of preparing to fight. General Miles exhibited endless patience. He was in conference with the chiefs and leading men all day long; he sent presents of tobacco and food to the Sioux who still hesitated about coming in. Visits to the agency by the people in these camps were encouraged, all weapons taken from the men when they entered the agency were scrupulously returned to them when they left. Slowly General Miles convinced the Sioux that he intended them no harm.

The military officers at Pine Ridge were now in hourly expectation of the final surrender, but the Sioux were still hesitating and the ghost dance leaders were very suspicious. On the fourteenth, Young-Man-Afraid-of-His-Horse brought in Little Wound with Little Hawk (Crazy Horse's uncle) and Crow Dog, now the most influential men among the ghost dancers, for another council. At this talk General Miles finally persuaded these men to bring in their people, and on the fifteenth they came streaming in, the movement starting early in the morning. By noon the whole country north of the agency was covered by moving parties of Sioux, all making for the agency. They streamed up the west bank of Big White Clay, most of the men on ponies, families in ancient farm wagons and afoot, and hundreds of ponies with bundles of tipi poles tied to their sides, the ends dragging on the ground behind them, with tipi covers and other property piled on the poles. The Oglalas formed a camp near the friendly camp west of the agency, while the Brûlés put up their tipis along the creek bottoms one-half mile north of the agency. By evening 742 yellowed and smoke-begrimed tipis had been put up, the huge camp extending for a mile along the creek, and in this camp were 3,000 to 3,500 Sioux who had given up. This time they had really given up, for General Brooke's troops that had formed a line far north of the agency now moved slowly southward on the heels of this mass of Sioux and formed a new line close around their camps on the north, west, and east. General Miles's own force closed in on them from the south, completing the encirclement. Asked for their weapons, the new arrivals gave up two hundred rifles. Big Road's band gave up nine guns, the chief claiming these were all his large band possessed. Unlike Colonel Forsyth, General Miles did not stir up these Sioux afresh by attempting to search them and their camps for hidden weapons. The horror of Wounded Knee was too fresh in every man's mind for the military to risk a repetition of that affair on a twenty times larger scale. The Sioux undoubtedly were concealing their best rifles; but what did it matter? They had no intention of fighting, and as the days passed and the Indians quieted down, more and more weapons were voluntarily given up.

In summing up his views on the ghost dance trouble, which he had personally observed at Pine Ridge, Brigadier General L. W. Colby of the Nebraska National Guard wrote: "This Indian war might be regarded as the result of a mistaken conception or misunderstanding of the Indian character and of the real situation and conditions on the reservations.... The general condition of things, however, which made such misunderstanding possible was the result of the Indian policy of the government."[5]

The men who had made that policy and forced it by pressure methods on the government—the brethren of the benevolent societies who styled themselves Indian Friends—met, as was their annual custom, in the parlor of the Riggs House in Washington on January 8, 1891, to adopt with a show of public debate and careful consideration the Indian policy for the coming year, which their leaders had decided on at Lake Mohonk. These leaders, meeting at Mohonk in the autumn of 1890, were in blissful ignorance that anything serious was wrong on the Sioux reservation. They were happy over their success in taking all of the surplus lands away from the Sioux in 1889 and were planning, now that they had the Sioux tied up (as they imagined) to put more

[5] *Nebraska Historical Society Transactions*, III, 170.

and stronger pressure on them and force them to progress more rapidly. Now in January in the Riggs Parlor, with the news of the slaughter at Wounded Knee fresh in their memories, the leaders among the brethren were in a peculiarly un-Christian state of mind: angry and spiteful, ready to utter violent accusations against any man whom they suspected of rolling this very large stone downhill and wrecking their neat arrangements on the Sioux reservation. They were as upset as hens that had nested in the brush on that reservation and had been suddenly blown off promising clutches of eggs by an outrageous wind. They were vindictively seeking scapegoats, and naturally they chose Red Cloud and the other old and almost superannuated chiefs, those wicked old nonprogressives, forgetting in their temper that these old men had all worked desperately through the early winter of 1890 to prevent bloodshed and that it was the younger men (the very class they were fond of describing as the hope of the Sioux) who had gone mad, taking up the ghost dance and carrying it on to its inevitable conclusion on the bloody field at Wounded Knee.

Listening to the reading of letters freshly come from the Sioux country, the brethren were aghast. The Reverend Tom Riggs (at Standing Rock) wrote that although the Yanktonais in the progressive camps along the Missouri were in their homes, all schools had closed and all work was at a standstill. The so-called nonprogressives had fled from their camps after the killing of Sitting Bull and were cowering with their families in snow-filled ravines, in hourly fear of being attacked by the troops. At Rosebud all the big Sioux settlements west of Little White River were abandoned, most of the people having fled to Pine Ridge to escape from the troops, burning their cabins before they left. The Rosebud progressives were huddled in tipis at the agency, under the protection of troops whom they did not trust. At Pine Ridge all the progressive settlements had been distant from the agency and the ghost dancers had wrecked them. The dancers had stolen most of the Indian livestock, and they had run off and eaten the government beef herd. Similar conditions prevailed at Cheyenne River where, as at Pine Ridge, all the Sioux were living in a concentration camp at the agency, guarded by troops. Through all the vast reservation those visible fruits of progress—cabin homes, fields under cultivation, and herds of stock—had been struck a dreadful blow, but worse than that was the news that those Sioux whom the Eastern "friends" of the Indians had fondly termed progressive had given up. They were, at least for the present, bitter and disillusioned. Their mainspring of action, their faith in the goodness and wisdom of the whites, had been broken. After all, they were Sioux by blood, and they had found Wounded Knee a very bitter cup to drain.

Being what they were, the brethren in the Riggs parlor were untroubled by the possibility that someone might be wrongheaded enough to point a finger in their direction and say, "There are the men and women mainly responsible for this outbreak among the Sioux." They sat in their comfortably padded chairs and worked off their temper on the nonprogressive Sioux, on Congress and the Harrison administration, and then their natural optimism began to bubble forth again and they fell to admiring the silver lining in this black cloud. After all, some said, the messiah craze and Wounded Knee were just the last kick of the wicked nonprogressives, and now that the Sioux had been taught this needed lesson the way was cleared for immediate and great progress. Commissioner of Indian Affairs Morgan agreed in that view. He said that there was no more reason for feeling despondent over Wounded Knee than over the Hay Market riot in Chicago—a curious comparison. He quoted Indian Office figures to prove how bright the Indian future should be, figures that one might fairly describe as deliberately doctored.

Senator Dawes, who was the principal support of the brethren and their cause in the Senate, now spoke, but not very cheerfully. He was angry because some of the speakers had laid the blame for Wounded Knee at the door of Congress. This he asserted was a false view. Congress had carried out all of the promises General Crook had made to these Indians in 1889, and the Sioux alone were at fault. He forgot to state that Congress had passed the necessary bills too late to stop the ghost dance troubles, and that it had acted then only under heavy pressure, General Miles having told everyone in Washington that the army could not end the

Sioux troubles unless Congress provided for the proper feeding of these Indians. Referring glumly to the bright future for the Indians which Commissioner Morgan and the other speakers had described with such optimism, Dawes said that he was in no humor for hunting rainbows. These plans the gentlemen had put forward would cost a mint of money and had no chance with a Congress that was disillusioned, angry, and bent on slashing appropriations. Indeed, Congress was considering once more—as it did in every Indian crisis—placing the tribes permanently under the War Department with army officers acting as Indian agents; and if that were done, where would all the bright plans the gentlemen had just been discussing end? The War Department would not touch these very expensive social experiments for Indians.

The brethren listened politely to Dawes, but they had their minds made up in advance and his words had no real effect. They waited until he sat down and then went on with their cut-and-dried program, passing a resolution approving their new plan of Indian policy.

Dawes had been correct in his forecast. Congress was in no mood for spending huge sums on Indian experiments. Then, in 1892, there was a political upset that put the Democrats in control and sent Grover Cleveland to the White House for a second term. The leaders of the Indian welfare groups were mainly Republicans; but their own party had refused to accept their Indian program, and now they hopefully took their new 1893 model plan to Cleveland. They also gave him the name of a gentleman of their own choice whom they wished to have appointed as commissioner of Indian affairs. Cleveland received the delegation as old friends and listened to them with attention, but a short time thereafter he appointed as head of the Indian Office a loyal Democrat whose very name was unknown to the leaders of these Indian welfare groups. The Indian Rights Association promptly tried to arouse the public by printing a pamphlet in which Cleveland's appointment of the new commissioner of Indian affairs was denounced as pure politics. That was just what it was, and the public did not care. The brethren after twenty-five years of Indian crusading had been frozen out, and again the public did not care.

At Pine Ridge the surrendered ghost dancers, bountifully fed and provided with warm new winter clothing, were permitted to quiet down in their camps under guard of the troops. While the Indian Friends in the East were angrily demanding wholesale arrests among these Indians and the punishment of all leaders, General Miles was scrupulously adhering to his promise that if the ghost dancers surrendered, they would not be badly treated. He was doing everything possible to reassure the Sioux and to bring them back to a normal condition of mind. He now selected delegations of chiefs from each of the big agencies and sent them to Washington for conferences at the Indian Office. At these meetings the Sioux gave their version of the recent troubles, and on the whole it was an honest and clear-cut narrative. They traced the beginning of the trouble back to the cutting of rations, to the drought and the destruction of their crops, to the taking of their lands in 1889, and to the government policy of trying to force them to progress rapidly, by every means, including cutting rations to starve them.

The leaders of the Indian Friend groups had been quite right in supposing that the military once in control among the Sioux would ignore most of their own program. General Miles would have nothing to do with the system, which he termed a forcing process. He ignored the long-established policy of pretending that chiefs were tyrants, standing in the way of progress, and all his negotiations with the Sioux were carried on through the chiefs. He shocked the Eastern Indian Friends and the missionaries by sending old Two Strike to Washington as the head of the Rosebud delegation—Two Strike who had been denounced as the principal leader of the Brûlé ghost dancers. He did this because the Brûlés themselves wanted to be represented by Two Strike. Red Cloud did not go to Washington. He who had made more journeys East than any Indian who had ever lived, was now in retirement. He had had enough of journeys, of councils, and of trouble. Short Bull, the Brûlé ghost dance prophet, with his lieutenant, Mash-the-Kettle, and a few others, were the only ghost dancers who were punished. Their punishment was a queer one. They were sent to Europe with Cody's Wild West show.

By the summer of 1891 the Sioux were back in the old rut, moving gently on in the slow evolution that constituted the only honest solution for their problems. The belief, expressed by Commissioner Morgan and the leaders of the Indian Friends, that Wounded Knee would prove a salutary lesson to the Sioux and would waken the conscience of the public to the need for greater efforts toward solving the Indian problem, had proven vain. The public had forgotten Wounded Knee within three months, and now in the summer of 1891 that field of useless carnage was carpeted with brilliant wild flowers amidst which crooked poles were dotted, stuck into the hard earth and daubed with Indian paint by the Sioux, to mark the spots where their people had fallen.

By this time the Indian Office officials were back in the rut too. The military agents at Rosebud and Cheyenne River had gone, and with no idea of anything better to do, the Indian Office had restored Agent Wright at Rosebud and Agent Perian P. Palmer at Cheyenne River. These gentlemen took up their routine exactly where they had had to drop it when the ghost dance troubles had blown them away in 1890. They both, indeed, were trying to keep up the old farming crusade, urging their Sioux to plant larger acreages than ever before, while the Dakota drought waited for August to show what it could accomplish in the way of crop destruction. McLaughlin at Standing Rock was carrying on his old program as if nothing had happened in 1890. At Pine Ridge, where many of the Brûlé ghost dancers of Rosebud had been permitted to settle permanently, it was considered wise to keep a military officer in charge for the present. Here the Sioux were attempting little farming, but a serious effort was being made to start them in earnest at cattle growing. The Sioux did not care particularly what work their white men set them to do. The thing that really mattered from their point of view was that the troubles of 1890 had compelled Congress to loosen the purse strings, and for the present at least they were receiving full rations and abundant supplies.

Thus the last stand of the Sioux as a tribe against the attempts of the government to make them over into imitation white people came to an end, with the Sioux very little changed and with the self-styled Indian Friends defeated in their effort to force these Indians suddenly forward along the way that they termed progress, defeated and, for the time being at least, out of favor in official quarters. On the whole the Sioux had come through this very trying period of transition from 1877 to 1891 very well. They had experienced many strange shifts and changes during the fourteen years; they had learned a little, mostly in the hard way, and they had taught the officials a little about Indians, those gentlemen learning—like the Sioux—in the hard way. The story of those years has in it much of foolishness, a good sprinkling of downright wickedness, some noble actions, and a goodly count of lessons hardly learned and, too often, quickly forgotten. That the Sioux had come through so well was mainly the result, one may believe, of their own good native qualities; for they were and are a good wholesome people, who rarely act in an unreasonable or violent manner except when driven to it by bad treatment.

Index

A Sioux Chronicle

is machine composed in 11-point Linotype Janson. The chapter titles are hand-set in the italic of the same face. Perpetua initials, designed by Eric Gill of England, appear inside a Linotype border at the chapter openings. A hand-drawn version of a rugged brush script by Johannes Boehland of Germany is used for the book title on the title page. The Indian war shield on the same page is a freely drawn rendering of this motif taken from a Sioux hide painting.